中国科学技术馆馆史书系
HISTORY BOOK SERIES OF CHINESE
SCIENCE AND TECHNOLOGY MUSEUM
中国科学技术馆 编著
COMPILED BY CHINA SCIENCE
AND TECHNOLOGY MUSEUM

THE TORCH OF SCIENCE
AND TECHNOLOGY IN ANCIENT CHINA

中国古代 科技之光

科学普及出版社
·北京·

图书在版编目（CIP）数据

中国古代科技之光：汉英对照 / 中国科学技术馆编著.
— 北京：科学普及出版社，2018.9
（中国科学技术馆馆史书系）
ISBN 978-7-110-09874-5

Ⅰ.① 中…　Ⅱ.① 中…　Ⅲ.① 技术史 – 中国 – 古代 – 图集
Ⅳ.① N092-64

中国版本图书馆CIP数据核字（2018）第185321号

策划编辑	郑洪炜
责任编辑	李　洁　陈　璐
装帧设计	中文天地
责任校对	杨京华
责任印制	马宇晨

出　　版	科学普及出版社
发　　行	中国科学技术出版社发行部
地　　址	北京市海淀区中关村南大街16号
邮　　编	100081
发行电话	010-62173865
投稿电话	010-63581070
网　　址	http://www.cspbooks.com.cn

开　　本	889mm×1194mm　1/16
字　　数	335千字
印　　张	15.25
版　　次	2018年9月第1版
印　　次	2018年9月第1次印刷
印　　刷	北京盛通印刷股份有限公司

书　　号	ISBN 978-7-110-09874-5 / N·107
定　　价	168.00元

《中国科学技术馆馆史书系》序

中国科学技术馆是我国唯一的国家级综合性科技馆，是实施科教兴国战略、人才强国战略和创新驱动发展战略，提高全民科学素质的大型科普基础设施。除了提供科学性、知识性、趣味性相结合的展览内容和科学教育，中国科学技术馆还承担了流动科技馆、科普大篷车、数字科技馆、农村中学科技馆等项目的管理和服务，自身服务于中国特色现代科技馆体系建设与科普事业发展的职责和任务不断拓展，促进理论研究、引领事业发展使命光荣、任重道远。

自1988年9月一期工程建成开放，中国科学技术馆已走过30个年头。30年中，中国科学技术馆经历了一期、二期和新馆三个阶段的建设发展；30年里，中国科学技术馆事业在迎接各种发展机遇、应对各种挑战中砥砺前行；30年来，几代中国科学技术馆人在学术研究、实践探索和开拓前进中积淀了宝贵的经验，积累了一批珍贵的历史文献、技术资料。

不忘初心，方得始终；以史为鉴，得失明知。中国科学技术馆策划、编著"中国科学技术馆馆史书系"丛书，旨在真实记载中国科学技术馆的发展历程，系统反映重要的展览展品内容，深度挖掘其不为人熟知的魅力，客观总结建设发展的经验，让中国科学技术馆的历史焕发出勃勃生机。丛书兼顾学术性、资料性和可读性，成熟一本，出版一本，力求质量，系列出版，自成体系。希望不仅能帮助读者了解中国科学技术馆事业发展历史，同时能对展览展品设计和科技教育工作提供启发、借鉴和帮助。

殷皓

中国科学技术馆馆长

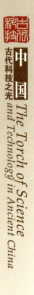

Preface to History Book Series of Chinese Science and Technology Museum

As the only national-level comprehensive science and technology museum in China, China Science and Technology Museum (CSTM) is a large-scale science popularization infrastructure for implementing the strategy of rejuvenating the country through science and education, the strategy of strengthening the country with talents, and the strategy of innovation-driven development and improving the scientific literacy of the whole people. In addition to providing the exhibition contents and science education that integrate scientific, informative and interesting features, CSTM also undertakes the management and service of mobile science museum, science circus wagon, digital science and technology museum, rural middle school science and technology museum projects. Its duties and tasks of serving the construction of the modern science and technology museum with Chinese characteristics and the development of science popularization keep continuously expanding, and its mission of promoting the theoretical research and carrying forward the arduous and glorious career also has a long way to go.

Since the Phase I works was completed and opened in September 1988, CSTM has been in business for 30 years. In the past 30 years, CSTM has experienced three phases of construction, i.e. Phase I, II and new venue. In the past 30 years, CSTM has been forging ahead by welcoming various development opportunities and coping with various challenges. Over 30 years, several generations of CSTM personnel have accumulated valuable experiences in academic research, practical exploration and pioneering & advancing, and also deposited a large number of precious historical documents and technical materials.

Never forget why you started, and your mission can be accomplished. Learning from history, and you can be fully aware of gain and loss. The "Book Series about the History of CSTM" plotted and compiled by CSTM aims to truly record the development history of CSTM, systematically reflect the important exhibition contents, deeply explore their unknown charms, objectively summarize the experiences of construction and development, and make the history of CSTM show great vitality. For these book series, an overall consideration will be given to the academic, informative and readable nature, the publication frequency shall depend on the maturity level of each of book series. In this process, we will strive for the quality guarantee, and pursue the goal of serial publication and self-established system. It is hoped to help readers understand the development history of CSTM career, and provide inspiration, reference and help for exhibition & exhibits design and scientific education work.

Yin Hao

Director-General of CSTM

August 2018

Preface 序

3

序言

中国是一个有着悠久历史的文明古国，华夏先民的勤劳智慧创造了灿烂辉煌的中国古代科技文化，中华民族为推动世界文明进步作出了巨大的贡献。16世纪中期以前，中国的科学技术一直居于世界前列。早在3000多年前的商代，甲骨文中就出现了日食、月食现象的记载；在西汉时期（前206—25）就已经发明了造纸术；在唐代（618—907）发明了黑火药，并在10世纪中叶用于军事；在宋代（960—1279）指南针已用于航海，并发明了活字印刷术。

英国著名科学史学家李约瑟认为："中国在公元3世纪到13世纪之间始终保持着一个西方望尘莫及的科学知识水平"。另一位同样著名的英国学者罗伯特·坦普尔在其著作《中国的天才人物》中指出："'近代世界'赖以形成的基础发明和发现，其中可能有一半以上均滥觞于中国""假如不是从中国引进了诸如船舵、指南针和多桅杆这些船舶和航海方面的改良技术，欧洲的航海大发现永远都不可能进行，哥伦布就不可能远航美洲，欧洲人也就不可能建立殖民帝国……如果不是从中国引进了纸张和印刷术，欧洲人靠手去誊写书籍的时代可能会漫长得多，文化的传播也不可能会像今天这样广泛……"。令人欣慰的是，随着时间的推移，并在中外学者赓续不已的努力之下，世界上越来越多不带偏见的人已经逐渐认识到并承认华夏先民对人类共同的科技遗产所作出的无与伦比的贡献。

及至近代，中国在封建社会体系的禁锢下，科学技术的发展受到严重阻滞。相较之下，西方社会经历了两次工业革命、资产阶级革命以及近代科学革命，极大地提高了社会生产力，科学技术得以迅猛发展，迅速超越了中国。今天，科学技术的日新月异，预示着一个新的、激烈竞争时代的到来，而对于中国，机遇和挑战并存。曾经的辉煌只能代表过去，实现中华民族伟大复兴的"中国梦"靠我们一步步努力前行，一代代接续奋斗。

正是本着这种精神，我们编撰了这本《中国古代科技之光》图文集。该书精选

汇集了中国科学技术馆历年来所开发的关于中国古代科技、发现和发明的大量展品的图片和文字资料，浓墨重彩地再现了古代中国在天文、指南针与航海、火药与火器、造纸术、印刷术、青铜冶铸、机械发明、纺织刺绣、井盐开采、建筑、陶瓷诸多领域的辉煌成就。以上这些成就，中国科学技术馆以互动的形式，在其"华夏之光"主题展厅以及"中国：7000年的发现""中国：知识的摇篮"和"中国古代机械展"等系列中国古代传统科技巡回展览中，择其精要，予以展现。

谨以此书纪念众多中国古代伟大的科学家和能工巧匠，他们非凡的发明和发现，为近代科学技术的形成打下了坚实的基础。与此同时，我们也以此书向王振铎、刘仙洲、陆敬严、李约瑟、坦普尔等中外学者和科学史学家致敬，正是得益于他们多年来的不懈努力，才使得中国古代科学文明不至湮没无闻，并最终能够走向世界；使得中国的故事能够用全球皆可接受的国际表达得以传播；使得"今天的技术世界是东西方的共同产物"这个真理能够深入人心。我们更加希望，这本书的出版发行，能够有助于构筑起一座跨文化的桥梁，使得横亘在东西方之间的"心理天堑"变为通途。

是为序。

徐延豪

第十三届全国人民代表大会常务委员会委员

中国科学技术协会

党组副书记、副主席、书记处书记

PREFACE

China is an ancient civilization with a long-standing history. Through their hard work and intelligence within millenniums, the ancient Chinese created a splendid culture of science and technology, contributing enormously to the progress of the world civilization. Before the middle of the 16th century, China had been at the forefront of the world in terms of science and technology. Back more than 3,000 years ago in the Shang Dynasty, records of the solar and lunar eclipses could be found in the oracle bone inscriptions; during the Western Han Dynasty (206 B.C.–A.D. 25), papermaking was invented; the invention of black gunpowder came in the Tang Dynasty (A.D.618–A.D.907) and its use passed from alchemists to military men in the middle of the 10th century; during the Song Dynasty (A.D.960–A.D.1279), compass began to be used in maritime navigation and the technology of printing with movable types came into being.

Dr. Joseph Needham, an eminent British historian of science, believed that "the Chines succeeded in maintaining, between the 3rd and the 13ht centuries, a level of scientific knowledge unapproached in the West". In his famous book *The Genius of China,* Robert Temple, another equally famous British scholar, points out: "possibly more than half of the basic inventions and discoveries upon which the 'modern world' rests come from China", "without the importation from China of nautical and navigational improvements such as ships' rudders, the compass and multiple masts, the great European Voyages of Discovery could never have been undertaken. Columbus would not have sailed to America, and Europeans would never have established colonial empires... without the importation from China of paper and printing, Europe would have continued much longer to copy books by hand. Literacy would not have become so widespread...". As time goes by and thanks to the ceaseless efforts by Chinese and foreign scholars alike, more and more people worldwide without prejudice have gradually come to recognize and acknowledge the immense contributions made by the ancient Chinese to the shared scientific and technological heritage of humanity.

However, in the modern era and under the shackles of the feudalistic social system, the development of science and technology in China was gravely impeded. In contrast, the Western society experienced two industrial revolutions, as well as the Bourgeois Revolution and the Scientific Revolution, all of which greatly promoted the productivity growth and brought about the speedy advancement of science and technology. As a result, the West surpassed China in a rapid way. Nowadays, the rapid development of science and technology promises a new era of intense competition, and opportunities coexist with challenges for China. Past glories are bygones. For the great Chinese Dream of revitalizing the Chinese nation to turn into reality, we must march forward step by step and make steadfast efforts.

It is precisely in this spirit that we have compiled *The Torch of Science and Technology in Ancient China*. With selected photos and information materials of the numerous exhibits on the ancient Chinese science, discoveries and inventions developed by China Science and Technology Museum over the past decades, the book relates at length the brilliant achievements of ancient China in the fields of astronomy, compass & Voyage, gunpowder & firearms, papermaking, art of printing, bronze metallurgy, mechanical inventions, textile embroidery, well salt mining, architecture and ceramics, all of which are featured in an interactive way at the museum's "Glory of China" themed gallery, as well as at its serial traveling exhibitions celebrating the traditional science and technology in ancient China, i.e. *"China: 7,000 Years of Discovery"*, *"China: Cradle of Knowledge"* and *"The Exhibition on Ancient Chinese Mechanical Inventions"*.

We hereby dedicate this book to the many great scientists and skilled artisans in ancient China whose extraordinary inventions and discoveries helped lay a solid foundation for modern science and technology. With this book, we also salute the numerous Chinese and overseas scholars and historians of science, e.g. Wang Zhenduo, Liu Xianzhou, Lu Jingyan, Joseph Needham and Robert Temple. Together they worked unremittingly to prevent the ancient Chinese science and civilization from falling into oblivion and to finally help it go global; to effectively communicate the Chinese story with globally acceptable international expressions; and to drive home the truth that "the technological world of today is a product of both East and West." Above all, it is our sincere hope that the publication and distribution of this book will facilitate the building of a bridge of cross-cultural understanding that will help turn the so-called "mental chasm" between the East and the West into a thoroughfare.

Xu Yanhao

Member of the Standing Committee of the 13th National People's Congress

Deputy Secretary of the Party Leading Members Group

Vice President

Executive Secretary

China Association for Science and Technology (CAST)

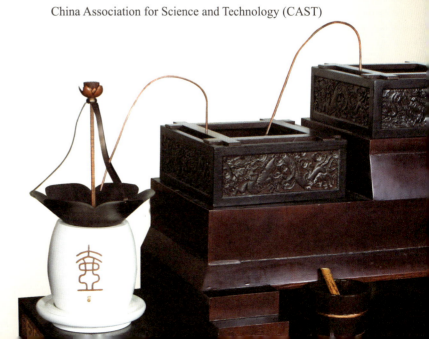

目录
Content

天

文

Astronomy

中国古代
科技之光
THE TORCH OF SCIENCE AND TECHNOLOGY IN ANCIENT CHINA

　　中国是世界上天文学发展最早的国家之一。早在新石器时代，先民们就开始了对日、月等天象的观察。据文献记载，远在4000多年前的尧帝就设有司天官专门从事观象授时。随着社会的进步，古代天文学得到迅速的发展。在天象观测方面，前16世纪，中国就有记载天象的文字，先民们相继留下的关于太阳黑子、彗星、流星、新星、日月五星的记事以及各种星图、星表，内容丰富，年代连续，其中许多还是世界上最早的记录。在天文学理论和天文仪器方面，他们创造了像浑天说这样颇有见识的宇宙观，发明了浑仪、简仪等光照后世的测天仪器。在历法方面，早在前16—前11世纪，中国就已经有了原始历法，经过不断改进完善沿袭至今。它既照顾了朔望月，又考虑了回归年，是别具一格的阴阳合历。

China is one of the first countries in the world to have started astronomical researches; observations of the sun, the moon and other astronomical phenomena date back to as early as the Neolithic Age. Documents indicate that as far back as some 4,000 years ago during the time of the legendary Emperor Yao, special officials were appointed to take charge of the observations of the astronomical phenomena. As the society progressed astronomy developed rapidly in ancient China. In the area of astronomical observations, there are written records dating from the 16th century B.C. about sunspots, comets, meteors, novas, the sun, the moon and five of the planets—Venues, Jupiter, Mercury, Mars and Saturn—as well as star catalogues, star charts etc. Rich in content and continuous in chronology, some of these records are the earliest of their kind in the world. In the fields of astronomical theory and instruments, the ancient Chinese established the famous theory of celestial sphere cosmology and invented such brilliant astronomical instruments as the armillary sphere and the simplified armilla. In calendrical science, the proto-calendar came into being in China between the 16th century and the 11th century B.C., which, having been improved and perfected through the ages, is still being used today. This calendar is a unique lunisolar calendar which takes into account the tropical years as well as the synodical months.

天象刻纹陶尊是古人用以祭祀日出、祈保丰收的礼器，出土于山东省莒县大汶口文化遗址，距今约 4500 年，目前对其刻纹大致有四种解释：①释为"旦"；②释为"灵"；③释为"日、月、山"；④释为原始部族的族徽。这些解释都与太阳相关，由此可以看出，早在原始社会人们就已经开始了对太阳的观测。

Dating from some 4,500 years ago, this sacrificial vessel was used by the ancient Chinese when they offered sacrifice to the sunrise and prayed for bumper harvests. At present there are generally four kinds of explanation about the meaning of the inscriptions on the vessel, namely: ① daybreak; ② spirit; ③ the sun, the moon and mountains; ④ the emblem of a primitive tribe. All the four explanations are related to the sun, indicating that the ancient Chinese started observation of the sun at a very early period.

漏壶是中国古代最早的水钟，一种非常重要的计时仪器。它由漏壶和箭型标尺杆组成。以壶盛水，通过水的流失计量时间，当壶内水面下降，箭尺也随之下沉，读出箭尺上的刻度即可知道具体的时刻。满城漏壶制作于公元前 113 年，1968 年于河北省满城的西汉刘胜墓出土。

The earliest known Chinese water clock, clepsydra was one of the most important timepieces used in ancient China. It was composed of an hourglass and a graduated arrow-shaped rod. As water runs out of the small hole in the bottom of the hourglass, the arrow-shaped measuring rod sinks on its float and shows the passage of time by its position. Excavated in 1968 from the tomb of Liu Sheng of the Western Han Dynasty in Mancheng of Hebei Province, the original of this clepsydra was made before 113 B.C. .

3 铜圭表
Bronze Gnomon

圭表是中国最古老、最简单的天文仪器，它由一根垂直的杆（表）以及一个带有刻度的标尺组成。整个装置南北向放置，通过测量正午时的表影长度可以推定节气和确定回归年的长度和季节。

下图为 1965 年在江苏省仪征县石碑村东汉墓出土的铜圭表，表长八寸（0.192 米），是标准表高的 1/10。圭表表面用枢轴连接，使用时将表竖立与圭垂直，平时可将表折入圭体内，成为一把尺子。

One of the oldest yet simplest Chinese astronomical instruments, it was composed of a vertical rod (the gnomon) set on level ground and a horizontal scale with marks at regular intervals lying due north-south. The length of tropical year, the seasons and solar periods could all be determined by measuring the gnomon's shadow cast on the horizontal scale at mid-day.

This is the replica of a gnomon excavated in 1965 from an East Han tomb, which was located in the Shipei village, Yizheng County, Jiangsu Province. Its vertical rod stands eight Chinese inches (about 0.192 meters) high, which is about one-tenth of the height of a standard gnomon. When folded it could also be used as a ruler. The gnomon is a folding designed portable tool. It can be easily deployed and gathered up.

4 地平式日晷
The Horizontal Sundial

地平式日晷是利用太阳投射的影子来测定时刻的中国古代计时仪器，其晷面放置于水平位置，晷面和晷针之间的夹角就是当地的地理纬度。下图为出土于内蒙古自治区呼和浩特市托克托城的汉代地平式日晷。

A time-measuring instrument used in ancient China, the horizontal sundial makes use of sun-cast shadows to determine the moment of time. The sundial is placed in a horizontal position, and the included angle between its scaled dial and gnomon is the local geographical latitude. The horizontal sundial shown in the picture was excavated from a Han tom in Togtoh City, Hohhot, the Inner Mongolia Autonomous Region.

　　赤道式日晷从地平式日晷演变而来，由晷针和晷面组成，是南宋（1127—1279）以来流行的利用日影方向的变化来计时的仪器。晷面平行于赤道面，晷针垂直于晷面，上端正指北天极，下端正指南天极，平行于地球自转轴。根据晷针在晷面投影的位置可以读出时刻，当晷针的影子指向正北方向时就是午时。

An instrument for observing solar time according to the direction of the sun's shadow projected from the gnomon, this equatorial sundial is typical of those produced in the Southern Song Dynasty (A.D. 1127–A.D. 1279). The sundial is composed of a bronze gnomon and a round-shaped stone slab which serves as a scaled dial. The centrally fixed gnomon was set vertical to the surface of the scaled dial that parallels the surface of the equator. The upper end of the gnomon points directly to the north pole whereas its lower end-points directly to the south pole, making it in parallel to the rotating axis of the earth. Under the sunlight when the sun's shadow cast by the gnomon points to due north, the time is 12 o'clock sharp at midday.

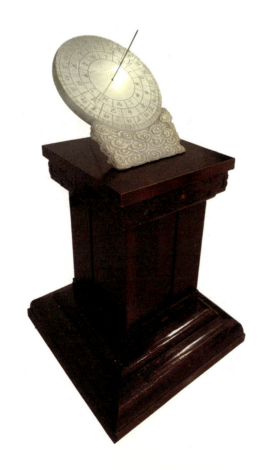

　　浑仪是中国古代用于测定天体坐标位置的天文仪器，最早发明于战国时期（前475—前221），后经历代天文学家改进，于10—11世纪定型。它可以测定天体的赤道坐标和地平坐标，还可以测定包括年、节气、月、日、时刻的时间"项目"。

　　下图为现存于紫金山天文台的明代（1368—1644）浑仪的1∶3比例缩小模型。

Armillary Sphere was an astronomical instrument used to measure the coordinates of celestial bodies. It was first invented in the in the Warring States period (475 B.C.–221B.C.), and attained its final form as we see today during the 10th-11th century. It can not only measure the equatorial and horizontal coordinates of celestial bodies but also measure the date and time such as years, solar terms, months, days and hours.

This is a 1∶3 scale model of the armillary Sphere which was made in the Ming Dynasty (A.D.1368–A.D.1644) and is now collected by the Zijinshan Observatory .

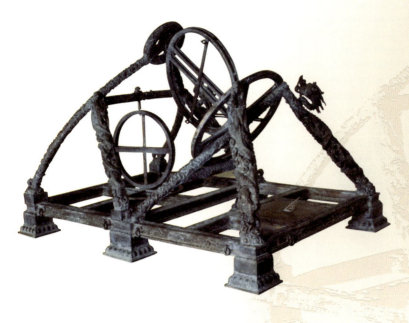

简仪是元代（1206—1368）天文学家郭守敬于1276年创制的测量天体赤道坐标与地平坐标以及真太阳时的仪器。

简仪由相互独立的赤道装置和地平装置组成，并以地球公转一周所需要的时间——365.25天为刻度，是当时世界最先进的天文仪器。简仪的赤道装置用于测量天体的赤道坐标。简仪的地平装置类似于现代的经纬仪，用于测量天体的地平方位以及地平高度。简仪的底座架中装有正方案，用来校正仪器的南北方向。在明制简仪中正方案改为日晷。

上图为现存于江苏省南京市紫金山天文台的明代正统二年至七年（1437—1442）制造的简仪的复制品。

Designed by the Yuan Dynasty (A.D.1206–A.D.1368) astronomer Guo Shoujing in 1276, this instrument was used to measure the equatorial and horizontal coordinates of celestial bodies, as well as the true solar time.

As the most advanced astronomical instrument when it was invented, the simplified armilla was composed of the mutually-independent equatorial mounting and horizontal mounting and graduated on the basis of 365.25 days, the length of time it takes for the earth to make one revolution around the sun. The equatorial mounting of the simplified armilla was used for determining the equatorial coordinates, the longitudes and horizontal coordinates of celestial bodies. Similar to contemporary altazimuth, the horizontal mounting of the simplified armilla was used for measuring the horizontal bearings and altitude of celestial bodies. In the base of the simplified armilla was a square board which was used for bearing calibrations.

This is a replica of the simplified armilla which was made in the period between the 2nd and the 7th years of the reign of Zhengtong (A.D.1437–A.D.1442) in the Ming Dynasty and is now collected by Purple Moantain Observatory in Nanjing Jiangsu Province of China.

浑象是古代用于演示天体视运动的仪器，相当于现代的天球仪。球面上绘有太阳、月亮、二十八宿以及赤道和黄道等。浑象靠水力带动，与天体同步运转，使人不受时间限制，可随时了解当时的天象。

An ancient version of the modern-day celestial globe, the celestial sphere was used for demonstrating the motion of celestial bodies. On the surface of the sphere were painted patterns of the sun, the moon, the Twenty-eight Mansions and other constellations, as well as the equator and the ecliptic. This hydraulic and automatic running device shows the same astronomical phenomena as in the sky, allowing people to make observations unrestrictedly.

水运仪象台是北宋天文学家苏颂和韩公廉于1086—1092年设计的以水为动力的天文钟，是现代天文台的雏形。其整体高约12米，宽约7米。台分三层：上层为观测天象的浑仪，中层为演示天象的浑象，下层为一套报时系统。这三部分用一套设计精巧的机械装置连接起来，利用流量稳定的水流带动，实现与天体运动同步的运转。底层的计时报时装置有160多个小木人，有些小木人举着木牌依次从木门内出现，用以显示时、刻，有些则定时敲击钟、鼓、铃、钲等乐器用以报时。这套报时装置还能够报昏、旦时刻以及夜间的更点。水运仪象台被认为是世界最早的天文钟。

Designed and built by Song Dynasty astronomers Su Song and Han Gonglian between the period of 1086–1092, this 12-meter high and 7-meter wide water-driven astronomical clock tower was, in essence, an observatory in the embryonic form. Elaborate gearing and transmission shafts connected the escapement respectively to an armillary sphere on the top platform, a celestial globe in the upper chamber and a time-telling system in the lower chamber of the tower, driving the three parts to rotate together. By controlling the uniformly flowing water, the central wheel was set to rotate in a certain direction and with equitime in order that the rotation of the armillary sphere and the celestial globe could remain in pace with the motion of celestial bodies. The time-telling system in the lower chamber was made up of an army of some 160 model men who emerged through doors, ringing bells, striking gongs or beating drums and carrying placards which announced the time, dusk and daybreak and even sounded the night watches. It is considered to be the world's earliest astronomical clock.

10

秤漏
Steelyard Clepsydra

秤漏是一种以秤称量流入容器中水的重量来计量时间的计时仪器，最早由北魏（386—534）道士李兰约于 450 年发明。隋唐时期（581—907）加以改进，成为官方的主要计时器。它的计时方式是以供水壶流到受水壶里的水的质量作为计时标准，以杆秤为显时系统。当流入受水壶的水为 1 升时，重量为 1 斤，时间为 1 刻（此处的"升"、"斤"均是中国古代度量衡单位，1 斤等于 16 两。"刻"为古刻，即 14.4 分钟）。

Invented in the Northern Wei Dynasty (A.D.386–A.D.534), the steelyard clepsydra was a timekeeper which measured time by weighing the weight or measuring the mass of water entering the water receiver from the water supplier. For instance,when the water entering the water receiver reached one Sheng which weights one Jin, it means one Ke had passed. ("Sheng" and "Jin" were unit of weights and measures in ancient China. One Jin equals 0.5 kilograms. Ke was quarter in ancient China, One Ke equals 14.4 minutes) A steelyard was used as the time telling system.

Thanks to the successive improvements made during the Sui and Tang Dynasties (A.D.581–A.D.907), the steelyard clepsydra gradually became the major official timekeeper.

莲花漏是宋人燕肃于1030年发明的一种根据水的流失来计量时间的计时仪器，由上匮、下匮和箭壶组成。通过虹吸管导流，水逐级流入箭壶之中。箭壶的浮箭随水量的增加而上升，根据箭上的刻度可以读出时间。燕肃在下匮的侧面开有一个分水孔，可以令多余的水由此流出，使下匮的水始终处于漫溢状态，从而保持水位的稳定，在很大程度上消除了水位变化对流量的影响，达到了准确计时的目的。

Invented by Yan Su of the Song Dynasty in 1030, the lotus clepsydra was a time-keeping instrument which use the lost amount of water to measure time. It consisted of the upper water container, the lower water container and a water-receiving vessel with a scaled floating arrow. By making use of the siphon principle, water flow in a constant speed from the upper container to the lower container and finally to the water-receiving vessel. As time went by, the rising water level inside the water-receiving vessel raised the floating arrow. By reading the marks on the scaled arrow, time could be measured and told.

12 铜壶滴漏
Bronze Clepsydra

铜壶滴漏是一种古代计时仪器，于元延祐三年（1316）制造。古代滴漏主要分单壶滴漏和多壶滴漏，铜壶滴漏属于多壶滴漏中最古老的一类，并且沿用至今。其通高2.64米，由日壶、月壶、星壶、受水壶四个带盖的壶组成，安放在一个类似台阶的架上。水从日壶中依次下滴，进入受水壶。壶中水位上升，木箭（标尺）随之上升，观其刻度，即知时间。

Made in the 3rd year of the reign of Yuanyou (A.D.1316) in the Yuan Dynasty, this kind of bronze clepsydras were used by the ancient Chinese for time-measuring purpose. There were mainly two categories of bronze clepsydras—the single-clepsydra style and the multiple-clepsydras style. The one shown here belongs to the latter category and is the oldest of its kind still in existence today. With a height of 2.64 meters, the set was composed of four vessels, namely the solar clepsydra, the lunar clepsydra, the stellar clepsydra and the water receiver. Each of them had a cover and all of them were placed on a stairs-style stand. From the solar clepsydra, water dripped down in proper order before finally entering the water receiver. As water kept dripping, the water level in the water receiver gradually rose, so did the wooden arrow (the measuring rod) floated inside it. By looking at the marks on the arrow, time could be measured and told.

大明殿灯漏是由元代天文学家郭守敬创制的计时仪器，因形似宫灯，又陈列于皇宫大明殿上，故称为"大明殿灯漏"。它利用水力带动机械装置报时，是世界上最早脱离了天文仪器的独立自鸣钟。

A timekeeping instrument designed by Guo Shoujing of the Yuan Dynasty, the original clepsydra resembled a huge place lantern and was once displayed in the Daming Palace, hence its name "the Daming Palace Lantern Clepsydra". It was a sophisticated mechanical device driven by hydraulic power, making it the world's first chime clock independent of any astronomical instrument.

14 苏州石刻天文图
Planisphere Carved on Stele

苏州石刻天文图是世界上现存最古老的石刻星图之一。其观测年代是北宋元丰年间（1078—1085），刻制于1247年。图总高约2.16米，宽约1.08米，上部为星图，下部为解释星图的铭文，星图自内规至外圈间刻28条不等距的宿度线，在图中可以看到银河以及1434颗恒星。

One of the world's oldest star charts of their kind still in existence today, the planisphere was drawn and then carved on a stone stele in 1274 on the basis of the observation results made during the reign of Yuanfeng (A.D.1078 – A.D.1085) in the Northern Song Dynasty. The planisphere was carved on a stele 2.16 meters in height and 1.08 meters in width. Its upper part is the planisphere and the lower part the explanatory inscriptions. From the innermost circle to the outermost circle of the planisphere, there are 28 straight lines of unequal distance representing equatorial coordinates. The entire planisphere shows 1,434 stars, it also indicates the bonds of the Milky Way.

历谱相当于后世的历书、日历。每日一简，分十二栏记录十二个月（如闰年，则是十三个月）的该日干支。汉元光元年历谱为编册横读式日谱，在特定的日期可以读出"冬至""立春""夏至"和"立秋"等节气。左图为 1972 年山东省临沂市银雀山二号汉墓出土的汉元光元年（前 134）历谱。该历谱对研究古代历法有重要的参考价值。

The 1st Year in the reign of Yuanguang of the Western Han Dynasty is 134 B.C. by the Gregorian calendar. Almanacs were tantamount to annual books or calendars of later ages. To be read horizontally, the almanacs of the Han Dynasty were mostly like that: the same date of all lunar months in one year was all on one bamboo slip, so there were 12 (or 13 if it was an intercalary year) dates on each slip. The dates were given the names of their heavenly stem and earthly branches. At the relevant date, it was noted with major solar terms such as "winter solstice" "the beginning of spring" "summer solstice" and "the beginning of autumn" etc. Excavated in 1972 from Tomb No. 2 of the Han tombs in Yinque Mountain at Linyi of Shandong Province, the almanac is of important reference value for researches into the calendars of remote antiquity.

中国古代科技之光 The Torch of Science and Technology in Ancient China

洛阳星图
Astronomical Chart Found in Luoyang

　　洛阳星图发现于河南省洛阳市北魏江阳王的墓顶，绘于北魏孝昌二年（526）。图直径7米，图中银河纵贯南北，波纹呈淡蓝色，清晰细致；星辰为小圆形，大小不一，计有300余颗。有些星用画线连起来，表示星座，最明显的是北斗七星，图中还刻画了许多其他单独的恒星作为陪衬。

Drawn in the 2nd year of the reign of Xiaochang in the Northern Wei Dynasty (A.D.526), the original astronomical chart was found inside the tomb of Marquis Jiangyang on the northern outskirts of Luoyang in Henan Province. In the center of this chart, which is about 7 meters in diameter, the Milky Way runs through from north to south and with clear and fine ripples in pale blue. The 300-plus stars in the chart are of varying sizes and are constelled with lines connecting them. Of all the stars, the seven stars in the Big Dipper are the most eye-catching. Also shown in the chart are the many other individual stars which serve as foils.

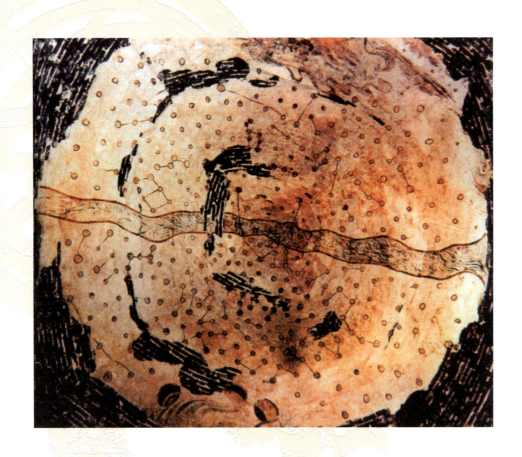

宣化星图于 1971 年发现于河北省张家口市宣化区的一座辽代墓中，星图绘于 1116 年，用于墓顶装饰，直径 2.17 米。星图中心嵌着一面直径为 35 厘米的铜镜，星宿围绕中心重瓣莲花作圆形分布。在莲花周围绘有九大圆圈，合为九星。在星图的第二层，按周天方位绘有二十八星宿，最外层是源于巴比伦的黄道十二宫，从中可以看出在天文领域内中外文化交流的迹象。

The original chart was found in 1971 inside a tomb of Liao Dynasty at the Xuanhua District of Zhangjiakou City in Hebei Province. This is a copy of an astronomical chart drawn in a tomb at Xuanhua in A.D. 1116. With a diameter of 2.17 meters, the chart was used as a decoration for the roof of the tomb. A bronze mirror 35 centimeters in diameter was inserted in the very center of the chart as a symbol of the center of the skies. The constellations are arranged in a circle surrounding the lotus flower in the center. Around the lotus flower are nine big circles. Together they represent nine stars. In the second ring of the chart were painted the twenty-eight lunar mansions. In the outermost ring were painted the twelve signs of the zodiac originating from Babylon, an indication of the cultural exchanges between China and the rest of the world in the area of astronomy.

二十四節氣

THE TWENTY-FOUR SOLAR TERMS

二十四节气是指中国农历中表示季节变迁的24个特定节令，是根据地球在黄道上的位置变化而制定的，每一个季节对应于地球在公转轨道上每运动15°所到达的位置，也就是太阳在黄道上的位置。二十四节气分为12个中气和12个节气，二者一一相间。

2016年11月30日，"二十四节气——中国人通过观察太阳周年运动而形成的时间知识体系及其实践"被正式列入联合国教科文组织人类非物质文化遗产代表作名录。在国际气象界，二十四节气被誉为"中国的第五大发明"。

The 24 solar terms refer to the 24 specific seasonal occasions expressing the changes of seasons in the Chinese lunar. They are determined based on the different positions of the earth on the ecliptic (the earth's orbit around the sun), with each term corresponding to a certain position, where the earth reaches after traveling each 15°. The terms consist of 12 pairs of major(sectional) and minor(middle) solar terms interlaced with each other.

In November 30, 2016, the 24 solar terms were formally listed in the representative work directory of intangible cultural heritage of UNESCO. In the international meteorological field, the 24 solar terms are reputed as "the fifth great invention of China".

18

二十四节气

The Twenty-Four Solar Terms

早在西汉时期（前206—25），二十四节气就已经得到了完善的发展。如今，二十四节气被认为是中国古代的第五大发明，被联合国教科文组织纳入人类非物质文化遗产名录。

当古代的中国人发现季节的变化与太阳有关时，他们便开始把已有的天文知识与太阳的运转联系起来，把太阳的年运行轨迹分成二十四个相等的部分，每个部分构成一个节气，每个节气相隔约十五天，每个月有两个节气，一年十二个月，刚好合成二十四节气。这一套与季节同步的历法诞生于中华文明摇篮——黄河流域。中国先民根据

天文学以及气候、降水和物
候序列变化的观测发明了这
一历法，它相当于一种"农
业年鉴"，成为规划农业生产
和指导日常生活的实用工具。
即使在今天，它仍然影响着
许多中国人的思维方式和行
为方式。

Completely developed by the time of the Western Han Dynasty (206B.C. – A.D.25) and now considered to be the country's fifth great invention, China's Twenty-Four Solar Terms has been added to UNESCO's List of Intangible Cultural Heritage of Humanity.

When the ancient Chinese discovered that the change of seasons had something to do with the sun, they began to apply their astronomical knowledge to determining the solar terms by dividing the sun's annual motion curve into 24 equal parts at intervals of 15 days, and each part has its own solar term. Summarized by Chinese ancestors based on observation of the sequential variations of astronomy, temperature, precipitation and phenology in the reaches of the Yellow River where the Chinese civilization was born, the season-synchronizing system has traditionally served as a kind of farmer's almanac of practical value in planning agricultural work and guiding daily routines. Even today, it still influences many Chinese people's ways of thinking and behavior.

指南针与航海

Compass and Voyage

指南针是中国古代四大发明之一。早在战国时期的史书中就记载了人类最早的指南工具——司南。中国于11世纪制成了人工充磁的指南针并开始用于航海，进而制成把磁针与方位盘连成一体的罗盘。同一时期，堪舆家发现了磁偏角，比西方哥伦布在航海中发现磁偏角早450多年。宋元时期，以指南针和罗盘为船舶导航，记录下航行中指南针在沿途岛屿、港口的指向，称为针位。各针位连贯起来形成针路，亦即航线。13世纪，指南针先后传入亚欧各国，对世界航海事业、经济、贸易、文化的发展起了巨大作用，推动了世界航海技术的发展。

The compass is one of the four great inventions of ancient China. In the ancient Chinese books dating back to the Warring States Period, there are numerous records of "*Sinan*", which literally means the "south governor" and is, in essence, the world's earliest compass. From such records we can see that as early as 11th century, "*Sinan*" was already invented in China and used by its people for indicating directions. In the Song Dynasty, a new kind of compass using the method of artificially-induced magnetism came into being and was widely used in sea-faring. It was on its basis that "*luopan*", a mariner's compass combining the magnetic needle and the directional pan was invented. Also in the Song Dynasty, magnetic declination was discovered by Chinese geomancer, whereas in the West it was not until 450 years later that Christopher Columbus made the same discovery. During the Song and the Yuan Dynasties, such compasses were used for navigating ships. With the aid of compass seafarers gradually explored and established ocean navigation routes which they called "needle routes." A certain navigation route or shipping course was connected together by many different "needle points" (compass points) which actually were seamen's records of the directions of islands and seaports passed on the voyages. During the 13th century, the use of magnetic compasses was gradually spread to various Asian and European countries, contributing tremendously to the developments of seafaring, economy, trade and culture throughout the world.

　　指南针是中国古代四大发明之一。公元前3世纪，中国人已经认识了磁石吸铁及指南现象，并制成了最早的磁性指南工具——司南。

　　在汉代王充的《论衡》一书中有"司南之杓，投之于地，其柢指南"的记载。司南由青铜盘与磁勺组成，方形青铜盘上刻有天干地支和八卦二十四方位。将磁勺放于铜盘上，转动勺把，待静止时，勺把指向南方。

Compass was one of the four great inventions of ancient China. In the 3rd century B.C. our ancestors successfully used natural lodestone to make the world's first compass—Sinan (the "south governor")

Detailed description about the shape and mechanism of Sinan can be found in *Lun Heng*, a book of the Eastern Han Dynasty by Wang Chong. According to the records in the book "the ladle-shaped lodestone of Sinan, when placed on the ground, would always come to rest with its handle pointing to the south." Sinan was carved in the shape of a ladle and balanced on a smooth bronze base-plate representing the heavens and the earth and engraved with 24 directional points. Spin the ladle, when it stops, the handle of the ladle always pointed to the south.

2 《梦溪笔谈》四种指南法
The Four Methods of Using a Magnetic Needle as Described in Mengxi Bitan

北宋科学家沈括于11世纪所著的《梦溪笔谈》中，记述了指南针的四种使用方法。分别为碗唇法、水浮法、指爪法、缕悬法。其中以缕悬法为最佳。

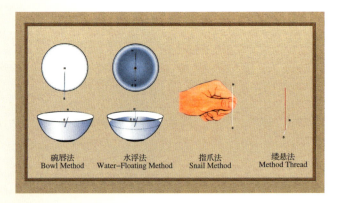

| 碗唇法 | 水浮法 | 指爪法 | 缕悬法 |
| Bowl Method | Water-Floating Method | Snail Method | Method Thread |

In his book *Mengxi Bitan* (*Dream Pool Essays*), which was written in the 11th century, Shen Kuo, a scientist of the Northern Song Dynasty, described in details the four methods of using a magnetic needle: balancing it on the rim of a bowl, floating in on water by passing it through a piece of rush, balancing it on a fingernail, suspending it by a single silk fiber (which is the best of the four methods).

3 缕悬法
The Fiber Suspension of a Magnetic Needle

缕悬法是《梦溪笔谈》中介绍的四种指南法之一，书中记载，"用新蚕茧的单丝作悬线用蜡黏合在磁针的中部，垂悬在一个木架上。木架下有方位盘，盘上刻有天干地支等，表示二十四方位，置于无风处，磁针静止时磁针两端指南北两向。"由于空气阻力小，磁针敏感性强，因此指向较精确。

Fiber suspension is one of the four methods of using a magnetic needle described by Shen Kuo in his book *Mengxi Bitan* (*Dream Pool Essays*). According to the book, the method consisted in "taking a single monofilament from a new cocoon and sticking it to the middle part of a magnetic needle with a tiny drop of wax, which is then suspended (on a wooden framework whose base is engraved with the heavenly stems and the earthly branches representing the 24 directional points) in a windless place." Due to the weak air resistance, this method is more accurate.

针碗是元代一种专门用水浮针的"王"字瓷碗。碗底绘有粗大三横，代表灯芯草，中间细竖代表磁针。使用时碗内注水，将插有灯芯草的磁针置于水面，浮针与碗内"王"字图像相对照以确定方向。

Used exclusively for the water floating method, this Yuan Dynasty porcelain compass bowl bears the Chinese character "王" inside the base of the bowl, with the three horizontal strokes of the character representing three pieces of rush and the vertical stroke in the center representing a magnetic needle. The compass bowl was used in this way: fill the bowl with water and passed a magnetic needle through a piece of rush so that it could float on that water. Then compared the floated needle with the diagram formed by the character "王" in order to determine the direction.

5

水罗经

Water-filled Compass

水罗经即水罗针，采用水浮法原理，由针碗演变而来。木质水罗经以整木雕成，盘面周围刻二十四方位，内中盛水，磁针横穿灯芯草，浮于水面以指示方向。

Known as floating needle compass, the water-filled compass was evolved from compass bowls. Using the floating needle principle, it consisted of a magnetic needle with a piece of rush floating in a water-filled azimuth bearing pan which was carved from an entire piece of wood and with 24 Chinese characters marked on its rim to indicate the compass points.

旱罗盘的盘面雕刻精细，标有二十四方位和八卦。旱罗盘中心有磁性指针，须把磁针重心置于高点，在阻力很小的情况下，磁性指针可以自由旋转，以指示方向。旱罗盘多用于营造宫殿、民居选宅时确定方向。

Exquisitely engraved with the 24 Chinese compass points and the 8 diagrams on its surface, the land compass pan has a magnetic needle placed in its center. In most cases, a land compass was used in the construction engineering of palaces; it was also used in choosing residential locations.

北宋《武经总要前集》中记载了指南鱼的制作及应用。将稍微带有凹度的薄铁片裁剪成鱼形，放在炭火炉中烧，待烧红后取出，鱼头向南，鱼尾向北，放入水中冷却后被磁化，漂浮于水面即可指示方向。

The making and application of compass fish were recorded in details in the book *Wujing Zongyao (Compendium of the Most Important Military Techniques)*. According to the records, thin strips of iron were cut into the shape of a fish, placed into a burning stove until they were red hot and then put into water to cool with the fish head pointing south and the fishtail pointing north. Contemporary science indicates that carbon steel, when heated to certain temperature and then cooled down rapidly in geomagnetic field, will result in phase change and become magnetized.

指南龟是根据南宋时期陈元靓的《事林广记》中的记载复原而成的。用木块雕刻成龟形，龟腹部中心安装一块磁石，龟尾部安放一根磁针，木龟的腹部挖一光滑的小孔，对准并放置于固定在木板上的尖状竹钉上转动，待静止时，首尾分别指示南北方向。

This model was reconstructed according to the description in the book *Shilin Guangji* by Chen Yuanliang of the Southern Song Dynasty. A lodestone was placed inside the belly of a wooden tortoise and a magnetic needle attached to its tail. The wooden tortoise revolved around a supporting bamboo stick. When it came to rest, its head and tail always pointed south and north respectively.

南宋时期（1127—1279）泉州古海船，于1974年在福建省泉州湾后渚港出土。其船型为尖头方尾尖底，这种设计使船能够在海浪中破浪前行。船底由两层木板构成，船舷用三层木板构成，全船底部隔为13个水密隔舱，增加了船只的横向强度和抗沉能力，也用于货物的存放。复原后的海船全船长34米，宽11米，排水量374.4吨，载重量200吨以上。在11—12世纪，中国的航海业与造船技术已经居于世界领先地位。

This ship was made in the Southern Song Dynasty (A.D.1127–A.D.1279), and was excavated at the Houzhu port of the Quanzhou Bay, Fujian Province in 1974. The original vessel had a pointed stem, a square stern and a sharp bottom, making it easy to cleave the waves and forge ahead. The bottom of the ship was built with double layers of wood board and the sides three layers. The entire ship was divided into thirteen water-tight compartments for the purpose of reinforcing the vessel's horizontal strength and increasing its anti-sinking capabilities. Designed to increase the stability of the ship, the compartments on its lower levels were used for cargoes. The ship was 34 meters long and 11 meters wide, with a displacement of 374.4 tons and a deadweight capacity of over 200 tons. It fully demonstrates that China was leading the world in the 11th and 12th centuries in terms of seafaring and shipbuilding technology.

泉州古海船的舱壁由铁夹固定，将船体分成13个水密舱。这种结构加强了船体的强度，改善了船舶的适航性和抗沉性。

The entire ship was divided into thirteen water-tight compartments by bulkheads fastened to the hull with iron cramps. With such a structure the hull was solidly joined and the strength of the ship reinforced, so was its seakeeping and anti-sinking qualities.

指南针与航海
Compass and Voyage

11 楼船
Towered Ship

　　我国早在春秋时期就有很多种大小不同，层数不一的楼船，汉代时，楼船盛行，其高大威猛的造型极具威慑力，可以有效对抗敌人的进攻。通常在战争中作为指挥船使用。它的建造和发展是高超造船技术的标志。

Dating back to as early as the Spring and Autumn Period, they were of varying sizes and decks. Due to its towering height, mammoth size and deterrent force, they were generally used as command ships in warfare. In the Han Dynasty, towered ships became prevalent; and their building and development was a symbol of the superior Chinese ship-building technologies at that time.

蒙冲是中国古代具有良好防护性能的进攻型快艇。它船形狭而长，航速快，专用于突击敌方船只。蒙冲有三个特点：①以生牛皮蒙背，具有良好的防御性能；②开弩窗矛穴，具有出击和还击敌船的作战能力；③以桨为动力，具有快速航行的性能。蒙冲是秦汉至唐代主要的攻击战船之一。

Noted for its well protective measures, fast speed, narrow and long shape, Mengchong was a kind of fast attack craft specially used in ancient China by waterborne troops for assaulting enemy vessels. Mengchong had three features: ① it used raw cowhide to reinforce its defensive capabilities; ② small windows were opened on both sides of the ship wherefrom to shoot arrows and throw spears, making it a warship capable of launching attacks and counterattacks against enemy ships; ③ multi-paddles were employed as propulsive tools to enable the ship travel fast. The Mengchong was one of the major warship from Qin、Han Dynasties to Tang Dynasties.

13
斗 舰
Combat Ship

斗舰是三国时期（220—280）至唐代（618—907）盛行的一种装备较先进的战船。船身两旁开有插桨用的孔，船体四周建有可用于侦察的女墙，女墙上皆有箭孔，用以攻击敌人。甲板上有棚，棚上也设有女墙，前后左右竖旗帜和金鼓，用来指挥作战，可壮声势。船尾高台上有士兵负责观察水面情况。前进的主要动力是船两侧数十只船桨的推动。

Prevalent in the long period from the Three Kingdoms Period (A.D.220 – A.D.280) to the Tang Dynasty (A.D.618 – A.D.907), the combat ship was a kind of man-of-war equipped with relatively advanced weapons. On both sides of its hull were holes for installing and maneuvering the paddles. Parapet walls were built around the hull of the ship wherefrom to spy on the enemies. At the same time, the arrow holes on the parapet walls could be used to launch arrow attacks against the enemy ships. The sheds on the deck also had parapet walls; and they were decorated with flags and metal drums that were used in commanding the battle fight, as well as in showing force. Soldiers were stationed on the high platform of the stern for reconnaissance purpose. Dozens of paddles were installed on both sides of the ship as propulsive tools.

　　明轮船是南北朝时期（420—589）发明的一种战船。明轮是一种船用推进器，装在船的两侧，形状如车轮，轮上有桨板，靠桨板拨水推动船只前进，因轮子大部分露出水面，故而称为"明轮"。在船内部，需要靠人力踩动桨轮轴，使轮轴带动轮子转动。宋代时（960—1279），最大的明轮船已达到长 36 丈（120 米）、宽 4 丈 1 尺（13.7 米），轮船从 2 轮发展到 4轮、8 轮、20 轮，甚至 32 轮。大型明轮船可装备弹射器用以投掷火药弹，是威力巨大的新型战船。明轮船的出现使船舶推进技术有了重大进展。

Paddle-wheel ship was invented as a kind of battleship during the period of the Division between the Northern and Southern Dynasties (A.D.420–A.D.589). As propellers of the ship, paddle wheels were installed on both sides of its hull. Inside the ship. Sailor stamped the paddle wheel shaft to drive the wheels. Such ships were called "open-wheel ships" in Chinese as their propeller wheels were largely exposed above the water surface. By the Song Dynasty (A.D.960–A.D.1279), the largest paddle wheeler had reached 120 meters in length and 13.7 meters in width. Accordingly, the number of propeller wheels increased from 2 in the beginning to 4, then to 8, to 20 and even to 32.

Large-scale paddle-wheel ships could be equipped with ejectors to catapult gunpowder projectiles, making such ships extremely powerful. The appearance of paddle-wheel ship was symbolic of the great advancement of ship propelling technology.

15 沙船
Sand Ship

沙船是中国四大古船之一，因其可在水浅多沙的航道上航行而得名，也称为"防沙平底船"。沙船起源于唐代（618—907），是一种大型运输船只。沙船的船头和船尾都呈方形，甲板宽敞，容积大，船舷较低。船体采用大梁拱，能够在强风大浪中快速前进。沙船的桅杆和帆很多，航行速度快，舵的面积大并且能够升降，可抗击七级风浪，在江河湖海皆可航行，适航性强。明代(1368—1644)远洋海运多用此船，运载量能够达到1200吨。

Originated from the Tang Dynasty (A.D.618–A.D.907) as a kind of large-scale transportation ship, the so-called "sand Ship" belongs to one of the four major types of ancient Chinese ships. It got its name because of the fact that it could sail on shallow water and along the sandy course. Sometimes it was also referred to as "sand-proof flat bottom ship". A typical sand ship had a square stem and stern with wide deck, huge capacity and relatively low shipboard. The hull of the ship had large camber so that it could cleave the waves fast and forge ahead in spite of strong wind and rushing tide. Its multiple masts and sails enabled it to travel fast. The ship was equipped with a large rudder that could be raised or lowered. Thanks to its strong seaworthiness, a sand ship could sail on rivers, lakes and seas. In the Ming Dynasty (A.D.1368–A.D.1644), sand ship, some of them with a loading capacity of 1,200 tons, were widely used for ocean shipping.

福船是中国四大古船之一，出现于宋代（960—1279），流行于福建、浙江沿海地区的一种尖底海船。福船可吃水四米，以行驶于南洋和远海著称，明代时（1368—1644）成为主要深海战舰。福船高大如楼，底尖上阔，首尾高昂，两侧有护板，甲板平坦，龙骨厚实，已采用水密隔舱技术，提高了船的安全性。

One of the four major types of ancient Chinese ships, "Fujian style ship" is a general term for the sharp-bottomed type of ships sailing along the coast of Fujian and Zhejiang Provinces in the ancient times. They first appeared in the Song Dynasty (A.D.960 – A.D.1279) and had a maximum draught of 4 meters, winning their fame for their frequent navigation to Southeast Asia and the open seas. During the Ming Dynasty (A.D.1368 – A.D.1644), Fujian style ships were used as the main oceangoing battleships. A typical Fujian style ship was towering like a building, had a sharp bottom and wide upper part, rising stem and stern, flat deck, reinforced keels, and with protective plates on both sides of its hull. The adoption of water-tight compartment technology greatly increased the safety of the ship.

17 郑和宝船
Zheng He's Ship

郑和是我国明代著名的航海家。1405—1433年，郑和船队曾受命七下西洋，历经亚、非30余国，行程达5万余千米，郑和所率领的最大舰队由60艘船舰组成，有27000名船员。

郑和宝船是下西洋船队中最大的海船，是当时世界上最大的木帆船。

宝船长44丈4尺（约140米），宽18丈（约57米），底尖上阔，船头昂，船尾高，建筑形式属于楼船，自底舱到甲板上，共分为5层。设9桅12帆，采用纵帆型布局、硬帆式结构，能够有效利用风能。宝船的排水量达到了2.85万吨。船体结构设置水密隔舱，提高了抗沉性和安全性，有利于分割舱段，分类载货，满足不同功能的使用要求。郑和宝船代表着当时中国极为先进的造船工艺水平。

By order of the emperor and with fleets of huge Ship, the famous Ming Dynasty navigator Zheng He made seven ocean voyages between 1405 and 1433 to Southeast Asia, India, Iran and the Arabian Peninsula, going as far as the Red Sea and the eastern coast of Africa and covering over 50,000 kilometers. The greatest of these fleets consisted of more than 60 vessels with a total crew of 27,000 men.

The biggest treasure boat is the largest of its kind in the world at that time. It was approximately 140 meters long and 57 meters wide. Built in the style of a towering ship, the treasure ship had a sharp bottom and a broad hull with raised stem and high stern. From bilge to the deck it has five layers; and with 9 masts, 12 sails, adopting fore-and-aft sail arrangement and hard-sail structure, it could effectively utilize wind power. The whole ship had a displacement of 28,500 tons. The water-tight compartment structure of its hull greatly increased the ship's anti-sinking ability and safety, making it easy to separate cabins, sort cargos and meet different functional requirements. In short, Zheng He's treasure ships represented China's extremely advanced shipbuilding technology at that time.

郑和下西洋，连接了中国到印度洋、红海、东非的航道，开辟了中国和世界航海史上的新纪元。郑和在远洋航海中将他所经过的各个国家的地理位置、航程远近和航道等都做了详细的记载，并绘制了《郑和航海图》图中标记了530多个地名，并绘有沿岸地形、山脉、岛屿、礁石、浅滩以及针路。《郑和航海图》收录在明代茅元仪编著的《武备志》中，被称为中国地图学史上最早的海图。

Zheng He's voyage to the west, linking China to India ocean, the Red Sea and East Africa. His voyage opened up a new era in the history of navigation between China and the world. Zheng He made detailed records of the geographical location, distance and channel of each country he passed in his voyage, and drew this nautical chart which marked more than 530 placenames and painted with coastal terrain, mountains, islands, reefs, shoals, and needles. Zheng He's nautical chart is recorded in *Wu Bei Zhi* compiled by Mao Yuan Yi in the Ming Dynasty. It is the earliest nautical chart in the history of Chinese cartography.

19

船尾舵
Stern Rudder

船尾舵是中国造船技术史上的一个重大发明。东汉时期（25—220）已有使用船尾舵桨的记载。船尾舵是设在船尾正中用以改变和保持船舶航向的设备，由舵柄、舵杆和舵叶三部分组成。船尾舵可以升降，在浅水区可将舵升起以免受损。在不需要改变航向时，可升起以减少阻力。古船上的船尾舵大致可分为不平衡舵与平衡舵两种。直到今天，船尾舵仍然是船舶航行的主要操纵工具。

10世纪，阿拉伯的航海者开始使用中国舵，12世纪末至13世纪初，船尾舵经由阿拉伯传入欧洲，船尾舵的发明是对世界造船技术的重大贡献。

Stern rudder is a major invention of China in the history of shipbuilding technology, and records of its application can date back to as early as the Eastern Han Dynasty (A.D.25–A.D.220). Installed in the middle of the stern of a ship and consisting of three parts, i.e. the tiller, the stock and the blade, stern rudders enabled people to steer their ships properly and with easiness. A stern rudder could be raised or lowered by rope tackle or chains. When entering shallow waters the rudder was often pulled up so that it would not be damaged; when it was unnecessary to change the course of the ship, the rudder was also raised in order to reduce resistance.

The stern rudders of ancient ships can be roughly divided into two types: the unbalanced rudder and the balanced rudder. Even today the rudder remains the major steering tool for ships in their navigation.

In the 10th Century, the navigators in Arabia began to use the Chinese rudder. From the end of 12th Century to the beginning of 13th Century, the stern rudder was introduced into Europe through Arabia. The invention of the stern rudder has made a great contribution to the world shipbuilding technology.

古代
科技

中国古代
科技之光

THE TORCH OF SCIENCE AND TECHNOLOGY IN ANCIENT CHINA

火药与火器

Gunpowder and Firearms

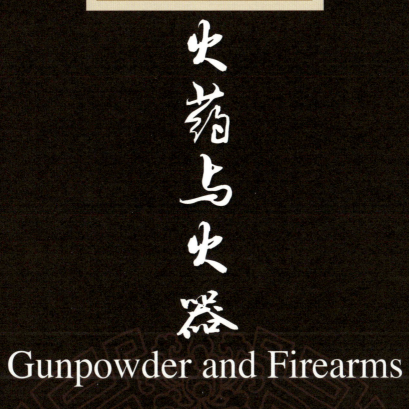

火药是中国古代四大发明之一。在 8 世纪末 9 世纪初，古代中国人在历代炼丹的基础上，摸索出以硫硝炭为基本成分的黑火药雏形配方。在 10 世纪，火药被制成火器。在 11 世纪曾公亮所著军事百科全书《武经总要》中记载了三种不同效用的火药配方。

火器的发展经历了燃烧爆炸性火器、喷射火器和管状射击火器三个阶段。明代军事经典著作《武备志》中记载了多种火箭类武器，包括并联火箭、多发火箭、返程火箭和二级火箭等。近代，火药在全世界范围内被用于和平事业，并取得了多方进展，在矿业、水利、交通、土木工程等领域得到广泛应用。

火药和火器于 13 世纪经由贸易渠道以及蒙古军的西征传入欧洲，对西方历史产生重大影响。

Gunpowder is one of the four great inventions of ancient China. During the late period of the 8th century and the early part of the 9th century, on the basis of alchemy practiced in the previous dynasties, our ancestors already came up with the rudimentary formula of making black gunpowder with sulphur, saltpeter and charcoal as the basic ingredients. In the 10th century gunpowder began to be used in making firearms. *Wujing Zongyao* (*Compendium of the Most Important Military Techniques*), a military encyclopedia written by Zeng Gongliang in the 11th century, contains records of the formulas for three different kinds of gunpowder.

The development of firearms experienced three different stages, namely, combustive and explosive firearms, injection firearms and tubular shooting firearms. *Wu Bei Zhi* (*Treatise on Armament Technology*), a classic military work of the Ming Dynasty, records a variety of rocket-type weapons, such as parallel rocket, multiple-firing rocket, retrievable rocket and two-stage rocket etc. In modern times gunpowder is used worldwide for peaceful purposes and great progress has been made in this regard. In such fields as mining, water conservancy, transportation and civil engineering etc. Gunpowder is also widely applied.

Gunpowder and firearms, which were to have a tremendous impact on Western history, came to Europe in the 13th century via trade routes and westward military campaigns of the Mongol armies.

《武备志》是明代茅元仪所著军事经典著作，也是研究我国古代火箭类武器的主要依据。

A classic military work by Mao Yuanyi of the Ming Dynasty, *Wu Bei Zhi* (*Treatise on Armament Technology*) is a major source central to research on the ancient Chinese rocket type firearms.

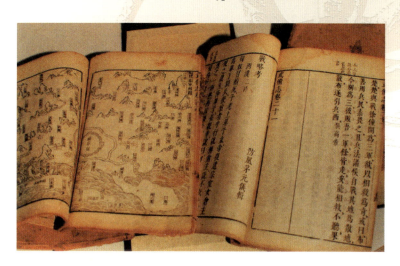

中国古代盛行炼丹术，唐代时达到鼎盛时期。9世纪的《真元妙道要略》书中记载，炼丹家将硫磺、雄黄、蜜和硝石混合起来烧炼，引起了爆炸和火灾。这些为求长生不老药而进行的尝试，虽然没有达到最初的目的，却启示人们发现了最早的制造火药的配方。

Alchemy was highly popular in ancient China; it was an experiment for elixirs of immortality. During the Tang, Dynasty, alchemy reached its heyday. According to *Zhen Yuan Miaodao Yaolue* (*Essentials of the Truly Original Methods*), a book written in the 9th century, alchemists mixed sulphur with realgar, honey and saltpeter and heated them together, resulting in explosions and fires. Such experiments, while failing to yield the originally desired results, led to the discovery of the first successful formula for making gunpowder.

中国早期黑火药的三种主要成分分别是硫、硝和木炭。8世纪的中唐时期，炼丹道士在炼丹过程中意识到硫、硝与木炭混合后的威力，从而发现了这种高度易燃易爆的混合物——火药。

Sulphur, saltpeter and charcoal constituted the three major ingredients of the early Chinese black gunpowder. By the 8th century in the mid-Tang Dynasty, Taoist alchemists, in their process of alchemical experiments, realized the potentialities of sulphur and saltpeter when combined with charcoal. This led to the discovery of a highly flammable and explosive mixture—gunpowder.

陶蒺藜为爆炸性火器，制造于13—14世纪，类似于现代的手榴弹和地雷。器内中空，可填装火药，作战时利用抛石机抛向敌方，也可布阵于地面阻拦敌兵前进。爆炸后，其尖锐的棱角可以对敌军造成重大伤亡。

Made in the period between the 13th–14th centuries, this explosive firearm was the ancient version of grenade and landmine. In battles, hollow pottery puncture vines were packed with gunpowder and hurled towards the enemies by means of catapults; or they could be laid on the ground to hinder the advancement of the enemy troops. When exploded their pointed edges and corners could inflict heavy casualties on the enemies.

石雷是明代（1368—1644）爆炸性火器，相当于现代的地雷。

A kind of explosive firearm of the Ming Dynasty (A.D.1368–A.D.1644), the stone mine was the equivalent to a modern day landmine.

6
元代铜火铳
Yuan Dynasty Copper Gun Barrel

火铳是由突火枪演变而来的管状射击火器。下图为1332年制造的元代火铳，以发射石制或铁制球形弹丸为主。一架大型火铳的威力巨大，有时会以"将军"对其命名。

Gun barrel is a kind of tubular shooting firearm evolved from firelock and blunderbuss. The one shown here was made in the Yuan Dynasty in 1332. Cannon with such gun barrel fired stone or metal cannon balls or grapeshot. A large gun barrel could be so powerful that it sometimes was referred to as "general".

7 抛石机
Catapult

火药发明后，人们认识到火药的军事威力，利用更早发明的采用杠杆原理的抛石机作为原始大炮，投掷火药包或石块来攻城守城。抛石机通常由多名士兵操作，火药包或石块的重量从不足 1 千克至几十千克不等，最大有效射程 500 米。

After gunpowder was invented and its military potential recognized, with the earlier invented catapults which made use of the lever principles, ancient Chinese developed a kind of primitive cannon for hurling gunpowder packs or stones in attacking or defending besieged cities. Normally, the catapults were operated by multiple soldiers, with the weight of powder packs or stones ranging from less than one kilogram to dozens of kilograms and the maximum effective range of 500 meters.

左图所示是根据《武备志》中记载的 17 世纪发明的虎头盾牌的复原品。虎头盾牌有攻守兼备之用。原物用木头制成，外面覆盖牛羊皮，上面画有虎头，虎头上方有圆形窥视孔，再用铁箍和钉子加固。该盾牌可抵御刀箭。盾牌两侧各有方孔，后藏有 8 支"火箭"，可通过方孔射出。亦可用长矛穿过上方圆孔进行近距离搏斗。

Used for both offensive and defensive purposes, the shield is a replica of the type invented in the 17th century and recorded in *Wu Bei Zhi* (*Treatise on Armament Technology*). The original was made of wood covered in cowhide or sheepskin on which a tiger's head was painted. Over the tiger's head was a round peephole. Reinforced with protective iron strips and nails, the shield could protect the soldier against the attacks of swords and arrows. Behind the rectangular holes on both sides of the shield was a hidden box containing eight "rocket arrows" that could be shot through the holes. Spears could also be used through the upper holes for hand-to-hand fighting if the enemy got that close.

一窝蜂是多发齐射火箭，形似箭袋，内装多达 32 支火箭，类似于近代的火箭炮。作战时，可肩扛亦可车运。古时一个营队配备几十甚至几百这样的火箭发射器。

右图为根据《武备志》记载复原的 17 世纪一窝蜂发射器。

A device which could shoot a number of fire arrows simultaneously, this weapon was shaped like a quiver and inside which were as many as 32 rocket arrows. It could be either carried on shoulders or mounted on a chariot. In ancient times, each battalion of troops was equipped with dozens or even hundreds of such launchers, which were equivalent to contemporary rocket guns.

Sourced from *Wu Bei Zhi* (*Treatise on Armament Technology*), this is a replica of an example used in the 17th century.

10 神火飞鸦
Rocket-Propelled Incendiary Called "Magic Flying Fiery Crow"

神火飞鸦的制作是用细竹篾、绵纸扎糊成乌鸦形状，内部填充火药，捆绑四支"起火"充当推进器。它是古代最早的带翼火箭，翼增加了飞行的稳定性和滑翔能力，借风力可以增加飞行高度和距离。鸦身内填充高度爆炸性火药，抵达目标后，内部火药被点燃爆炸。

下图是根据《武备志》记载复原的神火飞鸦，该武器主要用于在敌营的爆炸引火。

This weapon was woven of thin bamboo strips in the shape of a crow and with tissue paper pasted all over its surface. It was packed with gunpowder inside and propelled by four rocket arrows. As the ancient world's first winged rocket, its stability and gliding capability in the course of the flight were greatly enhanced; and with the help of wind power, it could gain a higher altitude and longer flight. The body of the "crow" was filled with a highly explosive gunpowder charge, which would be ignited by a fuse and explode when reaching the target.

Sourced from *Wu Bei Zhi* (*Treatise on Armament Technology*), this weapon was mainly used to set fire to an enemy's camp.

下图为复原的明代用独轮车装载火箭发射器的架火战车。车上装有6箱用于远程射击的火箭、2支短程小火铳、2支用于白刃格斗的长矛。

Vividly called "fiery chariot" by the Chinese, this is a reconstruction of a kind of mobile rocket launcher mounted on a wheelbarrow that was used in warfare in the Ming Dynasty. It was equipped with six boxes of rocket-arrows for use at long range, two small cannons for shorter range and two spears for hand-to-hand fighting if the enemy got that close.

12 火龙出水
Two-Stage Rocket Called "Fiery Dragon Emerging from Water"

　　火龙出水在古时用于水战，是世界上最早的二级火箭。龙身用茅竹制成，前后安装木刻的龙头、龙尾。将四支火药筒的引信汇总一起，以保证同步点燃，推动火龙飞行，这是第一级火箭。火龙腹内藏有数支小火箭。在飞行（通常在水面上几尺处作低平弹道飞行）过程中，待筒药将完时，自动引燃龙腹内的火箭。此时，从龙口里射出数只火箭，这是第二级火箭。

　　下图是根据明代茅元仪军事经典著作《武备志》中记载复原的火龙出水。

Used in naval engagements in the ancient times, this is the world's first two-stage rocket. The body was made from bamboo and the head and tail carved from wood. In the first stage of the launch, four gunpowder-filled tubes lashed to the body were ignited, propelling the dragon-shaped rocket forward. To ensure that all four powder tubes would be ignited simultaneously, their fuses were twisted together. A number of small rocket-arrows were hidden inside the body of the dragon. In the course of the dragon's flight (usually in a flat trajectory a few feet above the water) and just as the tubes were about to burn out, the rocket-arrows inside would be ignited by the tubes and launched from the dragon's mouth, which became the second stage of the launch.

The replica was made according to *Wu Bei Zhi* (*Treatise on Armament Technology*), a classic military work by Mao Yuanyi of the Ming Dynasty.

古代
科技

中国古代
科技之光

THE TORCH OF SCIENCE AND TECHNOLOGY IN ANCIENT CHINA

造纸术

Papermaking

　　造纸术是中国古代四大发明之一。早在西汉时期，我国的劳动人民就发明了造纸术。东汉元兴元年（105），蔡伦改进民间造纸术，以树皮、麻头、破布、废渔网为原料，成功地制造出大批量适于书写的纸张，从而促进了造纸术的发展和推广。纸，逐渐取代了简帛，成为最佳文字记载载体。造纸术在人类文明发展史上是一次划时代的革命。后来，中国的造纸术先后传入朝鲜、日本、阿拉伯国家和欧洲大陆。

Invented by the Chinese people in the Western Han Dynasty, papermaking is one of the four great inventions of ancient China. In A.D. 105, the first year of the reign of Yuanxing in the Eastern Han Dynasty and by making improvements on the papermaking techniques prevailed among the people, Cai Lun successfully produced a large quantity of paper suitable for writing out of tree bark, linen rags and worn-out fishing nets, thus greatly facilitating the development and promotion of the papermaking techniques. The appearance of paper as the incomparable and only best carrier of written records gradually overshadowed all previous writing materials such as bamboo slips and silk. Its invention was an epoch-making revolution in the history of human civilization. In the ensuing centuries after its invention, the Chinese papermaking technology was successively introduced to Korea, Japan, the Arabian countries and continent of Europe.

3000 多年前的殷商时期，人们把文字刻写在龟甲和兽骨上用于占卜记事，后人将这种文字称为甲骨文。甲骨文是目前已知的中国最早的成熟文字，通过甲骨文可以看到大量早期汉字发展的信息。

The earliest known mature written records in China date back more than 3,000 years, these are the oracle bone inscriptions, the accounts of divination incised on flat animal bones or tortoise shells by order of the rulers of the later Shang Dynasty. A great deal of information about the development of early Chinese characters has been gathered from such inscriptions.

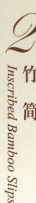

2 竹简
Inscribed Bamboo Slips

　　竹木简册是中国最古的书籍形制，起源于商代（约前1600—前1046）。单独的竹片称为简，木片称为牍。将若干根简用丝绳和牛皮带编缀在一起叫册。从先秦到两汉时期，大量的政府文件、著作都是写在竹木简册上的。直到4世纪初，竹木简册才被完全废止。

　　下图为1972年在山东省银雀山汉墓出土的竹简《孙膑兵法》。

As early as in the Shang Dynasty (approximately 1600 B.C.–1046 B.C.), bamboo slips (called Jian in Chinese) were bound together into what is known as the earliest books or Jian Ce. These narrow slips of bamboo on which records and documents were written or inscribed were strung together with silk cords or strips of cowhide. Wooden slips (called mu du in Chinese) were also used, chiefly for government documents, announcements and correspondence. It was not until the 4th century that this form of books was completely abolished.

Shown here is the replica of *Sun Bin's Art of War* inscribed on bamboo slips. The original was excavated in 1972 from a Han tomb at Yinque Mountain in Shandong Province, China.

古人将文字写在丝织品上，称为帛书。从公元前 5 世纪战国时期开始，在此后很长时期内，帛书同竹木简册一样是主要的书写材料。由于当时丝绸产量有限，成本高，所以帛书未被广泛使用，但它的发明促使人们寻求更轻便、更便宜的书写材料。

下图为 1972 年在湖南省长沙市马王堆汉墓出土的帛书《老子》残片。

During the Warring States Period in the 5th Century B.C., the Chinese book took a new form. The silk book, being made of soft pliable material, was rolled instead of being strung together like the inscribed bamboo slips. Though never been widely used because of the limited production of silk at that time and hence its high cost, its introduction nevertheless spurred the search for a lighter and cheaper material.

The displayed exhibit was reproduced on the basis of the fragments of the ancient silk book *Laozi* (*Laotzu*), which was excavated in 1972 from a Han tomb in Changsha, Hunan Province.

4 金 文
Bronze Inscriptions

　　商、周时期出现了各类制作精美的青铜器。这些青铜器除作为盛放酒、食的器具以及祭祀礼器外，皇室贵族们也会将一些描述重要事件、文件、宴会和成就的文字铸造在这些青铜器上以留存后世。这些文字就是金文。

During the Shang and Zhou Dynasties, there appeared a great variety of elegantly fashioned bronze ware. Besides using these as wine, food and ritual vessels, royalty and aristocrats had important events, documents, feasts and achievements etc. cast on them to keep for posterity. These are what would later be known as "bronze inscriptions".

1973 年在中国西北部的甘肃省金塔县金关出土了两张西汉麻纸（居延金关纸）。其中一张色泽白净，质地薄而均匀，而另一张则含有少量细麻螺纹，呈深黄色且较粗糙。经分析，是由大麻纤维制成的。

In 1973, two pieces of Western Han Dynasty hemp paper were excavated from Jinguan of Jinta County in Northwest China's Gansu Province. One of the pieces is of white colour and thin and even, whereas the other one, containing a small amount of ends of fine hemp thread, is of dark yellow colour and relatively rough. By analysis, it was made from hemp fiber.

造纸术
Papermaking

马圈湾纸是西汉麻纸的一种，于1979年在甘肃省敦煌西北马圈湾汉烽燧遗址出土。纸呈白色，质地细匀，纤维分布较均匀。

A kind of Western Han Dynasty hemp paper, it is of white colour and fine quality. The fiber distribution of the paper is rather even. It was found in 1979 at a Han Dynasty beacon fire site at Majuanwan, which is to the northwest of Dunhuang in Gansu Province.

旱滩坡纸是东汉麻纸，纸上有较大的汉隶墨迹，以麻为原料，质量很好，几乎没有纤维丝束和筛网的痕迹。旱滩坡纸于1974年出土于甘肃省武威县旱滩坡汉墓。

A kind of hemp paper made during the Eastern Han Dynasty, the paper bears Chinese characters written in the Han Dynasty style official script. With hemp as its raw material, the paper is of intense quality and hardly has any fiber tow and meshes. The original was excavated in 1974 from a Han tomb at Hantanpo of Wuwei County in Northwest China's Gansu Province.

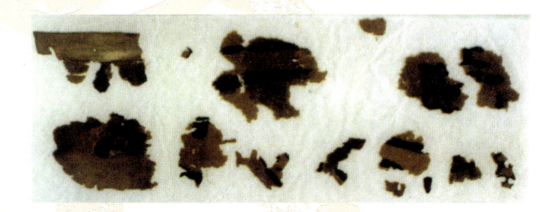

蔡伦，东汉(25—220)宦官，因发明造纸术有功而被封为龙亭侯。蔡伦曾任负责监造各种御用器物的尚方令，与全国的能工巧匠有着广泛的联系。东汉元兴元年（105），他总结前人造纸经验，带领工匠用树皮、麻头、破布、旧渔网为原料，成功地制造了一批适宜书写的植物纤维纸，这种纸很快取代竹简、帛书成为主要的书写材料。蔡伦的发明对造纸技术的发展与纸的推广使用起到了重要作用。为纪念蔡伦的卓越贡献，后人将这种纸称为"蔡侯纸"。

Known in China as the inventor of paper, Cai Lun lived in the Eastern Han Dynasty (A.D.25 – A.D.220) and was granted the title of Marquis of Longting. As an inspector in charge of the imperial workshops, he had wide contacts with artisans and workmen all over the country. After carefully summing up the experience he gleaned from them and with his own improvements, using tree bark, linen, rags and worn-out fishing nets, Cai Lun succeeded in producing a batch of paper made of plant fiber which was suitable for writing and presented them to the imperial court in A.D.105, the first year of the reign of Yuanxing. Because of the improved quality, the application of his paper soon became wide-spread and it gradually replaced bamboo slips and silk as the writing material. The later generations called such paper "Marquis Cai's Paper" in recognition of his contributions to the improvement of the papermaking technology.

"汉代造纸工艺流程图"形象地再现了两汉时期（前202—220）的造纸术：首先是将麻头、破布等原料用水浸泡，使之润胀，再用斧头切碎，用水洗涤，然后用草木灰水浸透并蒸煮。通过碱液蒸煮，原料中的木素、果胶、色素、油脂等杂质进一步被去除。用清水漂洗后，加以春捣。捣碎后的旧纤维用水配成悬浮浆液，再用漏水纸模捞取纸浆。经脱水、干燥后即成为纸张。

The chart vividly recreates the technological processes of papermaking prevailing in the Han Dynasty (202 B.C.–A.D.220): First of all, raw materials such as remnants of hemp, linen rags etc. were soaked in water to become swollen, then chopped with axe into tiny pieces. Secondly the chopped materials were washed in water and boiled with plant ash. After alkalization, lignin, pectin, pigment, grease and other impurities would be removed from the materials, which then would be rinsed with clean water and pounded into a loose pulp. After water and plant gums had been added, the pulpy solution was strained through a fine mesh screen on which the pulp dried to become sheets of paper.

《天工开物》是明代宋应星所著的一部技术百科全书，于1637年第一次出版。《天工开物》"杀青篇"中绘有五幅白描图，这些图表现了竹纸生产的五个重要工序：①斩竹漂塘；②煮楻足火；③荡料入帘；④覆帘压纸；⑤透火焙干。

Tiangong Kaiwu (*The Exploits of Nature's Works*) is a comprehensive encyclopedia on traditional technologies and practical skills written by Song Yingxing of the Ming Dynasty and first published in 1637. In this book there are five drawings showing the five important steps in the technological procedure for making bamboo paper: ① cutting bamboo and washing it in a pond; ② boiling pulp; ③ straining the pulpy solution through a fine mesh screen; ④ pressing it dry to form a sheet of paper; ⑤ drying the sheets on a heated wall.

斩竹漂塘
凡造竹纸事出南方，"登山砍伐（竹），截断五七尺长，放于本山开塘一口，注水其中漂浸……至百日之外，加功槌洗，去粗壳与青皮"。

煮楻图（蒸煮）
《天工开物》
卷中221页图
217页"（槌洗所得）竹穰形同苎麻样，用上好石灰化汁涂浆，入楻桶下煮，火以八日八夜为率"，"凡煮竹锅用四尺径者""（楻桶）围丈五尺，径四尺余。"

荡料入帘
《天工开物》
卷中222页图
218页"（匠
人）两手持帘，
入水荡起竹麻，
入于帘内荡，厚
薄由人手法。轻
荡则薄，重荡则
厚。"

覆帘压纸
《天工开物》
卷中223页图
228页"覆帘
落纸于板上，累
积于万张，数满
则上以板压；俏
绳入棍，如榨酒
法，使水气净尽
流干。"

火墙焙纸（烘纸）
《天工开物》卷中224、225页图。
218页"凡焙纸先以土砖砌成夹巷，
下以砖盖，巷地面数块以往即空一砖。
火薪从头穴烧发，火气从砖隙透，巷外
砖尽热，贴上焙干，揭起成帙"。

　　自纸发明之后，随着造纸术的发展，造纸原料的种类也不断增多。最早的造纸原料是大麻和苎麻，用其造出来的纸称为麻纸。东汉时期，蔡伦开辟了一种新的造纸来源——树皮，以树皮为原料造出的纸称为皮纸。魏晋南北朝时期又出现了以青藤皮为原料的藤纸。唐以后，以竹子为原料的竹纸兴起，并且迅速发展，很快取代麻纸、藤纸。从宋元到明清时期一直居于主导地位。除此之外，草也是造纸原料之一。

　　下图为造纸原料——楮树皮、竹、黄麻、桑树皮和龙须草。

After the invention of paper, the techniques of papermaking kept developing; and with the constant advancement of papermaking technology, the scope of raw materials used in papermaking steadily increased. Hemp and ramie were first used to make paper known as hemp paper. In the Eastern Han Dynasty, Cai Lun found a new source of papermaking material—tree bark—thus resulting in rough bast fiber paper. Then rattan paper made of vines appeared during the Wei, Jin and the Northern and Southern Dynasties. After the Tang Dynasty, hemp paper and rattan paper were replaced by bamboo paper, which was prevalent in the Song, Yuan, Ming and Qing Dynasties. In addition, grass in due time became one of the raw materials used in papermaking.

Shown here are some of the materials used in papermaking: paper mulberry bark, bamboo, jute, mulberry bark, and eulaliopsis binate (Chinese alpine rush).

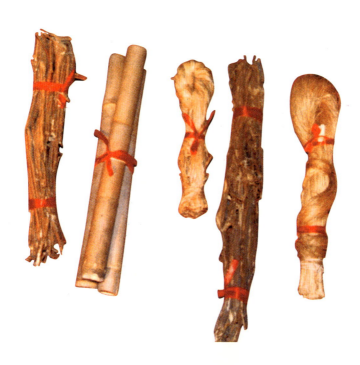

造纸术
Papermaking

12 宣纸
Xuan Paper

　　宣纸是出产于安徽省泾县，以青檀树皮为主要原料的纸。最迟出现于唐代。因其纸质洁白、柔韧、韵墨性好、吸墨性强，适于书画，故深受诗人、学者、画家和书法家的青睐。

　　下图为清代的两种宣纸制品。

Produced in Jingxian County of Xuancheng, a city in southeastern Anhui Province, Xuan paper, a particularly supple, thin, fine-grained paper, is often used for Chinese ink and wash and calligraphy. Appeared no later than the Tang Dynasty, it was especially coveted by poets, scholars, painters and calligraphers because of its outstanding absorbency. These qualities are attributable to the kind of tree bark (Pteroceltis bark) and rice straw used and to its sophisticated manufacturing processes.

Shown here are two kinds of Xuan paper made in the Qing Dynasty.

玉版宣
Strong White Xuan Paper

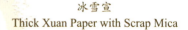

冰雪宣
Thick Xuan Paper with Scrap Mica

13

金粟山藏经纸

Jinsushan Scripture Paper

右图为宋代名纸，其上有朱印"金粟山藏经纸"，原物藏于浙江省海盐县金粟寺。这种纸的纸质优良，明清时期被人们收藏或用于装潢珍贵的书画。

Made in the Song Dynasty from mulberry bark, the paper bears the printing "Jinsushan Scripture Paper". The original was kept by the Jinli Temple in the Zhejiang Province which collected the best paper of the Song Dynasty. Noted for being elegant, semitransparent, anti-moth and water-proof, this kind of paper was regarded as a premium collection and often used to decorate precious calligraphy works and paintings during the Ming and Qing Dynasties.

14

洒金纸

Gold-Flecked Paper

洒金纸起源于唐宋时期，在明清时期逐渐盛行，是一种极为珍贵和精细的纸张。洒金是用金银粉在彩纸上描出各种花鸟蜂蝶等形象，或用黏性物在纸上绘出各种图案，再洒上金银粉末，使纸面光泽绚丽。

左图为清代洒金纸。

Originating from the Tang and Song Dynasties and prevailing during the Ming and Qing Dynasties, gold-flecked paper is an extremely precious and exquisite type of paper. Usually, it was made of bark and coloured with famille rose. Gold and silver powder was stuck to the surface of the paper and the whole calendered with wax. There are variations on this type of paper. For the most precious type, various patterns such as that of flowers, birds, bees and butterflies were painted on colored paper with either gold and silver powder or viscous substance before they were sprinkled with gold and silver powder, making the surface of the paper shining and gorgeous.

造纸术
Papermaking

15
清代对红纸
Qing Dynasty Red Paper for Writing Antithetical Couplets

对红纸是经过染色加工处理的清代古纸，用于写对联。

Dating back to the Qing Dynasty, this kind of ancient paper was dyed red and used for writing antithetical couplets.

16
入潢纸
Yellow-dyed Paper

入潢纸是经过黄檗溶液（黄柏树皮提取液）染黄，以起到防虫作用的纸。有时还会用蜡进行研光处理。

Dyed yellow with phellodendron amurense (bark of the amur cork tree) solution, this kind of paper is moth prooping. Sometimes it was calendered with wax.

暗花纸起源于唐代，用于抄纸的细帘上有刻花，在抄纸时会在湿纸浆上印上暗花。当拿纸对着光时就能清楚地看到发亮的线纹或图案。

Tracing backing to the Tang Dynasty, this kind of paper bears veiled patterns or designs formed by the papermaking screen while the pulp was still wet. When held against the light, such patterns or designs could be clearly seen.

造纸术在中国发明后，逐渐传到世界各地。4世纪前后传入朝鲜、越南。7世纪传往日本。8世纪中叶经中亚传到阿拉伯。12世纪西班牙人跟阿拉伯人学会了造纸，并于1150年在欧洲建造了第一个造纸厂，随后这项技术迅速在法国、意大利等其他欧洲国家传开。16世纪又从欧洲传入美洲与澳洲。19世纪时中国的造纸术已经在全球家喻户晓。造纸术的发明，为人类交流思想、传播文化作出了重大贡献，对世界文明的进步起到了巨大作用。

After its invention in China, the technique of papermaking was gradually introduced to different parts of the world. Around the 4th century, Chinese paper was introduced to Korea and Vietnam; and spread from Korea to Japan in the beginning of the 7th century. In the middle part of the 8th century, the technique of papermaking reached India and the Arabian countries through Central Asia. In the 12th century, Spain learned papermaking from the Arabs and in 1150 built Europe's first paper mill. Soon afterward the technique was taken to France, Italy and other European countries. In the 16th century, papermaking reached America, and sometime later Australia. By the 19th century, Chinese papermaking had become well-known throughout the world. The spread of the Chinese papermaking technique greatly promoted the development of the cultures of the world and played a tremendous role in the progress of the world civilization.

中国古代
科技之光

THE TORCH OF SCIENCE AND TECHNOLOGY IN ANCIENT CHINA

印刷术

Art of Printing

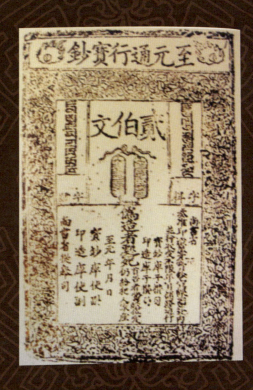

　　印刷术是中国古代四大发明之一。雕版印刷术发明于7世纪（隋末唐初），在宋代（960—1279）达到鼎盛。明清时期又在此基础上发展出精美的彩色套印技术。北宋庆历年间（1041—1048），毕昇发明了泥活字印刷术，这是印刷史上一次里程碑式的革命。在泥活字印刷术的基本原理上，后人又创制出木、铜、锡等不同材质的活字，并改进了排版材料和拣字方法，形成了铸字、排版、印刷等一套完整的印刷工艺。

　　约8世纪时，中国印刷术传入邻国朝鲜和日本，随后，向西传入波斯和阿拉伯地区，影响远至非洲和欧洲。

Printing as one of the four great inventions of ancient China was first invented in the 7th century, the last years of the Sui Dynasty and the outset of the Tang Dynasty. Also called "block printing", it reached its golden age during the Song Dynasty (A.D. 960−A.D.1279) as imperial patronage encouraged the publication of large numbers of books by the central and local governments. Between 1041 and 1048, on the basis of block printing, Bi Sheng invented movable type and was recorded by his contemporary Shen Kuo. Since then a complete printing process of typefounding, typesetting and printing came into being. What appeared soon afterwards were such techniques as process printing and woodblock water-color printing etc.

Somewhere around the 8th century, the Chinese printing techniques spread to neighboring Korea and Japan. Later they were transmitted westward to Persia and to the Arabian region. Their impact was felt as far away as Africa and Europe.

　　汉文字是世界上最古老的文字之一。印刷术发明之前，汉字字体先后经历了甲骨文、金文、大篆、小篆、隶书、楷书的深刻演变。纵观汉字的演变历史，是一个由圆到方、由繁至简的过程，字形也逐渐趋于优美。这一点对楷书来说尤其如此，楷书形成于三国时期，字体方正，横平竖直，更易写、易刻、易读，这也为雕版印刷的发明创造了重要的条件。

Chinese character is one of the oldest scripts in the world. Before the invention of printing technology, their calligraphic style had undergone profound changes in the long river of history, i.e. from jiaguwen (oracle bone inscriptions) to jinwen (inscriptions on bronze objects), then to dazhuan (the greater seal character current in the Zhou Dynasty), xiaozhuan (the lesser seal character adopted during the Qin Dynasty), lishu (the official script) and kaishu (the regular script). An overview of this evolution process shows that the form of Chinese characters gradually turned from round to square and from complexity to simplicity. Their shape also became more and more beautiful. This is particularly true of kaishu, which took its form during the Three Kingdoms Period. With straight horizontal and vertical strokes and a square form, kaishu was easy to write, to engrave and to read, thus creating an important condition for the invention of carved-block printing.

印刷术
Art of Printing

2 拓片 Rubbing

拓片是将石碑、青铜器等各种器物上的凹凸图文用纸、墨捶拓转印在纸上的工艺。这种以正写阴文复制正写文字的方法，成为中国印刷术发明的重要技术条件之一。隋代时拓印技术已很发达。

拓片技术直到今天仍在使用，可以用它将历史文物上的图文复制下来，同时又能保护原物的完好。这种工艺制作的副本常被书法家用作范本。

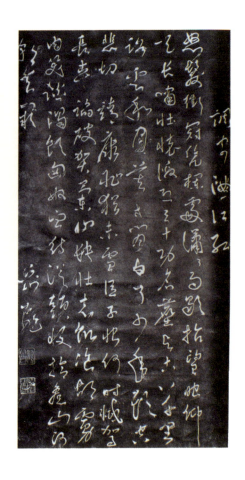

Rubbing is a printing technique employed to make copies of characters/designs carved in intaglio or in relief on various utensils. Such printing technique, which results in characters in observe on the basis of characters cut in intaglio, provided one of the pivotal technical conditions for the invention of printing technology in China. During the Sui Dynasty, rubbing technology already reached a very high level.

This rubbing technique is still used today as a means of making copies from cultural relics whilst preserving the original works. Copies made with such technique are often used as paragon by calligraphers.

下图展示了雕版印刷术的发明及发展过程。雕版印刷术发明于7世纪，它是将雕刻在平整木板上的文字、图像转印在纸张上的印刷技术。其工艺过程是：首先，按照固定的格式将手稿抄写于极薄的纸上，将稿纸反贴在枣木或硬梨木做成的木板上，然后雕刻成反向凸字。印刷时，将薄薄一层墨涂刷于木版表面，再覆上纸张，用刷子轻轻刷过，这样正向的字迹就印在了纸上。

随着雕版印刷术的发展，又逐渐衍生出铜版印刷、纸币印刷以及可以用多种颜色印刷出精美图画的彩色套印技术。

The following figure shows the invention and development process of block printing. Invented in the 7th century, carved-woodblock printing, also called "block printing" for short, was another Chinese invention and a major step toward faster printing methods. It was done by copying the manuscript according to a set format on thin sheets of paper. These were stuck face down on blocks of hard pear or jujube-tree wood and the reversed characters carved in relief. Printing was done by applying a thin coating of ink on the block, after which a sheet of paper was placed over it and stroked with a brush. A printed sheet with the characters right side up resulted.

Carved-woodblock printing also developed into an artistic medium, used for reproducing fine calligraphy and for creating delicate pictures, often in several colors.

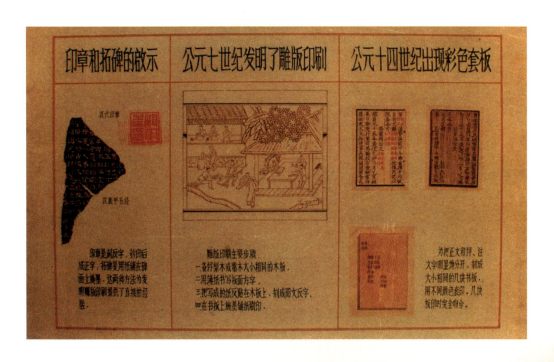

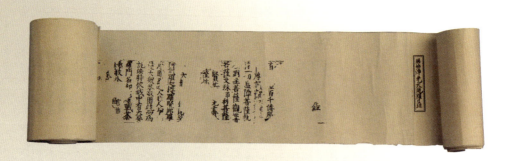

《无垢净光大陀罗尼经》是唐代印刷品，发现于韩国东南部庆州佛国寺释迦塔内。经卷纸幅共长 610 厘米，高 5.7 厘米，经文为楷书写经体，笔画道劲，字上有明显的木纹，刀法工整，是中国唐代武则天执政晚期的雕版印刷品。

A Tang Dynasty print, *the Great Dharani Sutra* was discovered in the Sakyamuni Pagoda of a Buddhist temple in the southeast part of the Republic of Korea. The entire volume of the sutra, written in kaishu or the regular script in a forceful manner, is 610 centimeters in length and 5.7 centimeters in height. The printed characters bear obvious wood grain and traces of carving. Tests indicate that this sutra was a print made with the block-printing technique during the last years of Empress Wu Zetian's reign of the Tang Dynasty.

《陀罗尼经》为唐代末期印刷品，1944 年发现于四川省成都市唐墓。采用薄茧纸，呈半透明状，上印有佛像和梵文咒语，边上还印有"唐成都府成都县龙池坊卞家"等字。为国内现存最早的印刷品之一。

A print done during the last years of the Tang Dynasty, *the Dharani sutra* was found inside a Tang tomb in Chengdu of Sichuan Province in 1944. The sutra was printed on semi-transparent thin silkworm paper. It bears pictures of Buddha and incantation written in Sanskrit. On the margin of the paper is a line which reads "Done by the Bian Family run Longchi Studio, Chengdu County, the Prefecture of Chengdu, Tang Dynasty". This is one of the oldest prints still in existence in China.

　　《金刚经》为唐代印刷品，20世纪初发现于甘肃省敦煌莫高窟石室。《金刚经》来自大乘佛教，崇尚从根本上放弃一切的尘世眷念。整幅经卷的卷首为插图，随后是经文，卷尾印有一行字，表明该经卷为咸通九年（868）四月十五日所印，这是现存最早的印有明确日期的雕版印刷品。

　　《金刚经》原件现藏于英国国家图书馆。

A Tang Dynasty print based on the Mahayana school of Buddhist teaching that advocates giving up extreme forms of worldly attachments, *the Diamond Sutra* was discovered inside a stone chamber of the Mogao Grottos in Dunhuang of Gansu Province in the beginning of the 20th century. The whole sutra is in the form of a scroll. It begins with an illustration and following with scripture. The end of the scroll bears a printed line which indicates it was done on the 15th day of the 4th month in the 9th year of the reign of Xiantong (A.D. 868), making it the oldest known and precisely dated woodblock print.

The original is now collected by the British Library.

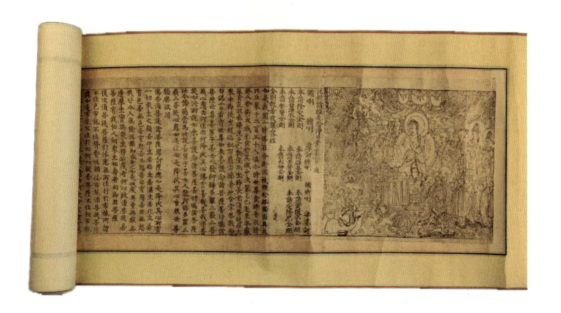

7

五代《大圣文殊师利菩萨像》

The Manjusri Bodhisattva Statue

《大圣文殊师利菩萨像》为五代时期（907—960）刻印品。

The Manjusri Bodhisattva Statue, a print which using the block printing technology, it dates from the Period of the Five Dynasties (A.D.907－A.D.960).

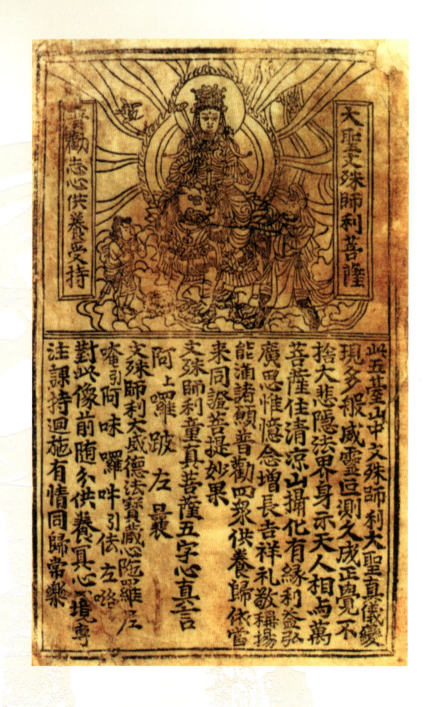

印刷术的一个主要功能是用来印刷纸币。南宋时期（1127—1279），被称作"会子"的纸币大量发行。下图为12世纪南宋都城临安（今杭州）印制纸币会子所用铜版的复制品。币面印有发行机关名称、面额及对首告伪造者的赏格等。

A major area of printing had to do with paper currency. By the time of the Southern Song Dynasty (A.D.1127–A.D.1279), paper banknotes called "huizi" were issued in large quantities. Shown here is the replica of a copper plate used for printing huizi at Lin'an (today's Hangzhou), capital of the Southern Song Dynasty in the 12th century. Printed on the bills were the name of the issuing agency, the value of the bill and the reward for the first person to report a counterfeiter.

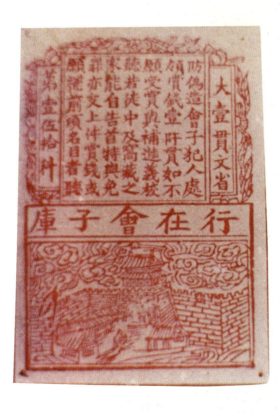

9

　　元代纸币通行于全国，包括偏远地区如新疆和西藏，甚至经由贸易活动传至东南亚、印度和波斯。

　　1287 年发行的元代纸币"至元通行宝钞"有 11 种票面，用铜版和专用纸张印制。下图为印制该种纸钞所用铜版的复制品。

In the Yuan Dynasty, paper banknotes circulated across the entire empire, including remote areas such as Xinjiang and Tibet. Through trading activities, such paper currency even spread to India, Persia and Europe.

In 1287, a new Yuan Dynasty currency called "zhiyuan tongxing baochao" was issued in eleven denominations. Copper plates and special paper were used for printing the new bills. Shown here is the replica of one of such copper plates.

上图为宋代山东省济南刘家功夫针铺用来印刷广告所用的铜版。铜版上刻有简洁文字和精致图案，是我国现存的古代第一个商业广告。

The copper plate was used by a Song Dynasty needle shop owned by the Liu Family of Jinan, Shandong Province, for printing advertisements promoting its products and services. The plate was engraved with a brief text and intricate designs. This was the first commercial advertisement in ancient China.

10
宋济南刘家功夫针铺广告铜版
Copper Plate Used for Printing Commercial Advertisements

印刷术
Art of Printing

11

元《无闻和尚金刚经注》 Monk Wuwen's Commentary on the Diamond Sutra

《无闻和尚金刚经注》为元代至正元年（1341）由湖北省江陵资福寺印制。首页上的书名，以及插图中的人物、桌子、云彩和灵芝，均用红色印刷，而其中的松树和经文则用黑色印刷。这是我国现存最早的朱墨双色套印本。

Dating from 1341, the 1st year of the reign of Zhizheng in the Yuan Dynasty, the book was printed by the Zifu Temple in Jiangling of Hubei Province. The title in the first page of the book, as well as the human figures, desks, clouds and Ganoderma lucidum depicted in its illustrations, were printed in red color, whereas some of the pine trees and scripture were printed in black color. This is the oldest extant book printed in double colors (red and black) with the process printing technique.

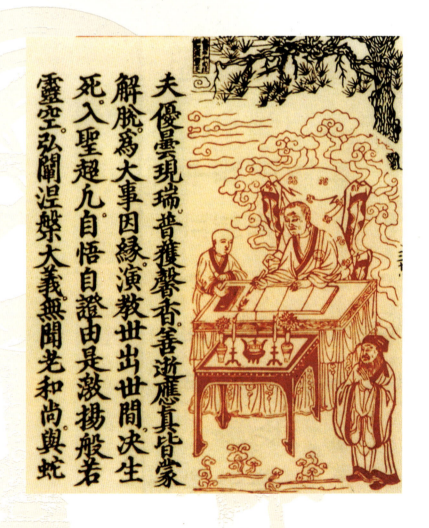

这本中国画彩色画谱于 1619—1626 年由明代出版家胡正言用其发明的饾版法印制而成，由十竹斋出版。饾版法就是根据图画中的不同颜色分色分版，刻成多块小木板，然后依次逐色套印，最后形成一幅完整的彩色画面。这种印刷方法清代以后称为木版水印。《十竹斋画谱》是这种印刷技术最早期的代表作品之一。

Done by the Ming Dynasty publisher Hu Zhengyan during 1619 to 1626 and with the dou ban technique (in which each plate is divided into a number of blocks according to the colors in the design. Then each block is impressed on paper with a different color to produce a complete picture) he invented, this colored copybook for traditional Chinese painting was produced by the Ten Bamboo Studio. This printing technique was called "woodblock watercolor printing" after the Qing Dynasty. This copybook was one of the earliest representative works using such printing technique.

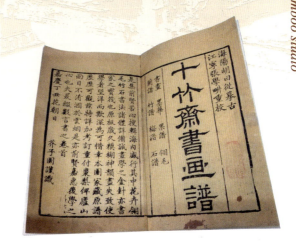

13 桃花坞年画
New Year Pictures Made at Taohuawu

苏州桃花坞木版年画已有约300年的历史。桃花坞年画由多种颜色套印而成，画面内容涉及传统戏曲、民俗民风和风景画等。桃花坞年画色彩鲜明，画面生动，深受中国南方人民的喜爱。

Taohuawu woodcut New Year pictures have a history of some 300 years. Printed with the multi-color process printing technique, the themes of the colored pictures produced at Taohuawu near Suzhou range from traditional operas, folklore, folkways and landscapes etc. Taohuawu New Year Pictures are especially favored by people of South China for their bright colors and vivid scenes.

14 杨柳青年画
New Year Pictures Made at Yangliuqing

杨柳青年画于明代崇祯年间（1628—1644）创于天津杨柳青镇。通过描绘传统戏曲、故事或仕女娃娃图，寓意美好幸福、喜庆吉祥。其制作方法是在木版套印的基础上再施以手工彩绘。时至今日，仍然是中国北方著名的民间艺术。

Yangliuqing, a village near Tianjin, began making New Year pictures in the Ming Dynasty during the reign of Chongzhen (A.D.1628–A.D.1644). Scenes from traditional operas and novels, as well as beautiful women and chubby children, are used to represent happiness, good luck and prosperity. The Yangliuqing pictures, still a well-known folk art today, particularly in North China, are given an added application of watercolor after printing.

中国古代科技之光

The Torch of Science and Technology in Ancient China

山东潍坊杨家埠年画创于清代（1616—1911）初年。年画多用红、黄、青、紫、绿五色，因色彩鲜艳而闻名。画面多反映人们渴求五谷丰登、家庭和睦的美好愿望以及孝亲爱幼、勤劳节俭的传统美德。

Yangjiabu, a village near Weifang of Shandong Province, began making New Year pictures in the early years of the Qing Dynasty (A.D.1616–A.D.1911). Noted for their bright colors, the Yangjiabu pictures often use the five colors of red, yellow, green, purple and blue. The content of the pictures often reflects people's desire for bumper harvest and harmonious family life, as well as the traditional virtues of filial devotion, hardworking and thriftiness.

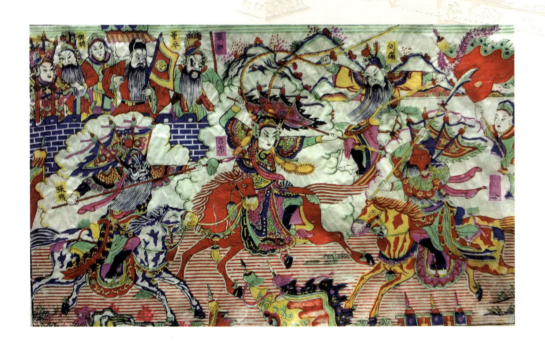

16

《活字印刷图》
Illustration of Printing with Movable Types

下图展示了制造活字所用材料的演变过程，包括泥活字、木活字、锡活字、铜活字和铅活字以及元代王祯发明的转轮排字盘。

This illustration shows the development of the materials used in making movable types, including clay, wood, tin, copper and lead. It also demonstrates the typesetting turntable which designed by Wang Zhen of the Yuan Dynasty.

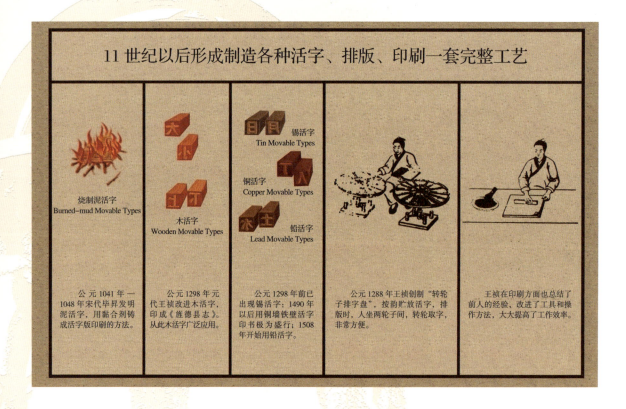

11 世纪以后形成制造各种活字、排版、印刷一套完整工艺

烧制泥活字
Burned-mud Movable Types

木活字
Wooden Movable Types

锡活字
Tin Movable Types

铜活字
Copper Movable Types

铅活字
Lead Movable Types

公元 1041 年—1048 年宋代毕昇发明泥活字，用黏合剂铸成活字版印刷的方法。

公元 1298 年元代王祯改进木活字，印成《旌德县志》。从此木活字广泛应用。

公元 1298 年前已出现锡活字；1490 年以后用铜墙铁壁活字印书极为盛行；1508 年开始用铅活字。

公元 1288 年王祯创制"转轮子排字盘"，按韵贮放活字，排版时，人坐两轮子间，转轮取字，非常方便。

王祯在印刷方面也总结了前人的经验，改进了工具和操作方法，大大提高了工作效率。

宋庆历年间（1041—1048），平民毕昇发明了活字印刷术。根据沈括《梦溪笔谈》记载：毕昇以胶泥为原料制作活字，以火烧硬。印刷前，在模里放上松香、蜡、纸灰等的混合物，上面一排排摆满活字，然后以火加热，待混合物熔化后，将活字压平，冷却后即可印刷。印完后，再加热铁板，等混合物熔解后，将活字取下，备下次再用。这种方法近似于现代印刷技术中排版的基本原理。

活字印刷术克服了雕版印刷费工、费时、雕版数量多存放不便等缺点，大大提高了工作效率，是印刷史上的一次重大革命。

Movable type printing was invented by a commoner named Bi Sheng during the reign of Qingli (A.D.1041–A.D.1048) of the Song Dynasty. This was recorded by his contemporary, Shen Kuo, in *Mengxi Bitan* (*Dream Pool Essays*). Bi Sheng fashioned types out of clay and hardened them by firing. Before printing, the types were placed in rows on a mixture of resin, wax and paper ash spread over a framed metal platen. Then the platen was heated to melt the mixture and the types pressed in. Once the platen had cooled it could be used as a printing plate. To release the types for further use, the platen needed only to be heated again. This method approximates the basic principles of type-setting in today's printing techniques.

Movable type printing overcame the shortcomings of block printing, which was usually laborintensive, time-consuming and required numerous carved blocks. It greatly increased the work efficiency and brought about a great revolution in the history of printing technology.

印刷术
Art of Printing

　　元代王祯是木活字印刷的改进者和发展者。1297—1298 年他设计制作了转轮排字盘，并发明了将活字按韵分类的方法。这种排字盘通常由两个轮盘组成，一个轮盘按韵存放活字字模，另一个则存放"杂字"。排版时两人合作，一人喊号，另一人坐于两轮中间转动字盘，方便地取出所需要的字模排入版内。

　　转轮排字盘的发明是排字技术的一项重大革新。

Wang Zhen of the Yuan Dynasty was the innovator and developer of wooden movable type printing. Between 1297 and 1298, he also designed the typesetting turntable and the method of classifying characters according to their pronunciation. Usually, such turntables were used in pairs, characters were arranged on one of the tables according to their pronunciation while the other table held "miscellaneous characters". When setting types one person would shout the characters to be set while the other person sitting in between the two turntables would conveniently pick out the desired types and put them into the tray by turning the turntables.

The invention of the typesetting turntable was a major innovation with regard to the typesetting technology.

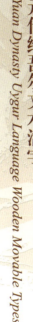

左图为元代（1206—1368）刻制的木活字，该木活字发现于甘肃省敦煌莫高窟内。上刻有回鹘文（维吾尔文），证明元代时期木活字印刷技术已广为传播。

Found inside the Mogao Grottos at Dunhuang of Gansu Province, the wooden movable types date back to the Yuan Dynasty (A.D.1206–A.D.1368). They were engraved in the Ouigour (Uygur) language and are evidence that during the Yuan Dynasty, the technique of printing with wooden movable types already spread widely.

清道光年间（1821—1850），安徽省泾县人翟金生仿效北宋毕昇造泥活字的方法，历时30年时间研究并制造了五种不同字号的泥活字约10万个。他用这些自制的泥活字，成功地印刷了自己的诗文集和翟氏宗谱等书。

下图为1962年发现于安徽省徽州的翟氏泥活字字模的复制品。

During the reign of Daoguang (A.D.1821–A.D.1850) of the Qing Dynasty, Zhai Jinsheng from Jingxian County of Anhui Province spent 30 years researching and making around 100,000 clay movable types, which were of five different specifications and made in imitation of the process used by Bi Sheng, the Northern Song Dynasty inventor of movable type printing. With those self-made types he succeeded in printing his own collection of poems and the Zhai Family genealogy.

Shown here are replicas of the clay movable types made by Zhai Jinsheng that were found at Huizhou of Anhui Province in 1962.

印刷术
Art of Printing

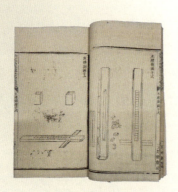

清乾隆三十八年至四十一年（1773—1776），清政府用枣木刻成约 25 万个木活字，印刷了两部重要的书籍，分别是《武英殿聚珍版丛书》和《钦定武英殿聚珍版程式》。《武英殿聚珍版丛书》共 134 种，2389 卷，包括《四库全书》及其他经典著作，几乎包括了经、史、子、集在内的历代重要著作，是我国历史上规模最大的一次木活字印书。而《钦定武英殿聚珍版程式》则详细记载了这次木活字印刷的情况，是研究我国古代印刷史的珍贵史料。

In the period between 1773 and 1776, or the 38th to 41st year of the reign of Qianlong in the Qing Dynasty, the Qing government ordered the making of some 250,000 jujube wood movable types for printing two significant books respectively entitled *Wuyingdian Juzhenban Congshu* (*Precious Collection of the Wuying Palace Series*) and *Qinding Wuyingdian Juzhenban Chengshi* (*The Procedure and Process for Printing Precious Collection of the Wuying Palace, Made by Imperial Order*). Composed of 134 categories and 2,389 volumes, the former includes *The Complete Library in Four Divisions* and other classical works, and is inclusive of almost all important works of the previous dynasties covering the four basic branches of Chinese literature (namely, classical works, historical works, philosophical works, and belles-lettres), making it the greatest printing endeavor by means of wooden movable types in the Chinese history; whereas the latter records in details the entire process of the printing activity. *Qinding Wuyingdian Juzhenban Chengshi* is a precious historical source for the study of the history of printing in ancient China.

中国古代
科技之光

THE TORCH OF SCIENCE AND TECHNOLOGY IN ANCIENT CHINA

青铜冶铸

Bronze Metallurgy

　　早在约 4000 年前，我们的祖先已能铸造出精美的青铜器。商周时期是我国青铜文化的鼎盛时期。公元前 14—公元前 11 世纪，古代中国人成功地铸造出重达 832.83 千克的后母戊鼎。在湖北省铜绿山发现的古矿遗址中，地下古矿井分布密集，至今仍堆积着 40 多万吨古代青铜炼渣。在长期的生产实践中，我国古代劳动人民总结出了一套完整的采矿、冶炼、制范、铸造技术；认识了青铜合金的性能，根据不同的需要采用不同的原料配比制成各类青铜器，包括礼器、兵器、乐器、生产工具和生活用品等，创造了光辉灿烂的青铜文化，为古代生产力的发展、文化艺术的提高和进入铁器时代铺平了道路。

Bronze wares symbolize a nation's etiquette and power. As early as 4,000 years ago our forefathers were already able to make bronzes of fine and exquisite quality rarely equaled by other bronze-age civilizations. During the period between the 14th century B.C. and the 11th century B.C., spanning the Later Shang Dynasty and the early part of the Zhou Dynasty, bronze casting reached its peak, as marked by the successful cast of the famous Hou Mu Wu Ding, a bronze quadripod weighing 832.83 kilograms. At the site of the Tonglushan ancient copper mine which was discovered in Central China's Hubei Province, more than 400,000 tons of bronze slag left over from ancient times are still piled up there today. Under the ground of the mine place is a densely dispersed network of ancient mine shafts and tunnels.

Over a long period of production practice the Chinese people developed a complete technological process of mining, smelting, mold-making and casting; realized the properties of bronze alloy; and, according to different demands, were able to use different alloys to produce various kinds of bronze wares, including sacrificial vessels, weapons, musical instruments, production tools and utilitarian bronzes. The mastery of bronze casting technology by the Chinese led to the creation of a splendid bronze culture and paved the way for the increase of productivity, the further development of culture and art, and above all, China's entry into the Iron Age.

铸造工艺 *Casting Techniques*

1

泥范法
Clay Mold Casting Technique

　　泥范法出现于商代早期，至商代中期达到鼎盛。古代工匠利用这种方法创造出了后母戊鼎、四羊方尊这样的旷世珍品。

Clay mold casting technique. It first appeared in the early period of the Shang Dynasty and culminated in the mid-Shang Dynasty. With this technique, the ancient craftsmen created unparalleled masterpieces such as the Hou Mu Wu quadripod and the Si Yang Fang Zun wine vessel.

2

叠铸法
Stacked-molds Casting Technique

　　叠铸法的出现和广泛应用，使生产率提高、成本降低。这种古代铸造工艺一直沿用至今。

Stacked-molds casting technique. Its appearance and widespread application resulted in the growth of productivity and the reduction of costs. Even today, this ancient casting technique is still in use in China.

青铜冶铸
Bronze Metallurgy

3 失蜡法
The "Lost-wax" Casting Technique

失蜡法起源于前5世纪的战国时期，在现代工业中仍在使用。失蜡法的工艺流程为：首先用蜡按铸件的样子制成铸模，然后放入箱中，铸模表面敷以多层黏土，干燥后用火烧，使黏土变硬，同时将蜡化去，形成空腔铸范，浇入熔化的青铜溶液，待青铜熔液凝固冷却后，砸掉黏土外壳，就得到一件和最早的蜡模一模一样的青铜铸件了。

The "Lost-wax" casting technique. First developed during the Warring States period in the 5th century B.C. and still used in modern industry, this casting technique with expendable molds involves the following processes: First, a wax model was made as a perfect image of the desired bronze object. This was then placed in a box and covered with several layers of clay. When this had dried it was fired to harden the clay and melt out the original wax model, leaving behind a perfect mold. Molten bronze was then poured into the space left by the wax model. When the bronze had cooled the clay mold was smashed off, revealing a casting which matched the wax model in every respect.

车马器 *Horses and Chariots devices*

1
铜车軎
Bronze Wei

这组铜车軎为西周（前1046—前771）时期的车马器，原物出土于河南省平顶山应国墓。它是用以加固车轴的车构件，通常套在车轴的两端，以防车轮脱出，一般与车辖配合使用。

Dating from the Western Zhou Dynasty (1046 B.C.–771B.C.), Wei was a kind of chariot implement used to reinforce the axle of a chariot. It was fastened at both ends of the axle of the chariot to prevent the wheels from coming off. Usually, it was used together with a linchpin. The original was excavated from a tomb of the State of Ying located at Pingdingshan of Henan Province.

2
车马杂器一组
Bronze Chariot Implements & Horse Harness

这组车马杂器，是西周时期的铜器。原物出土于河南省三门峡虢国墓。

Dating from the Western Zhou Dynasty, the original pieces of this group of chariot implements and horse harness were excavated from a tomb of the State of Guo located at Snamenxia, Henan Province.

青铜冶铸
Bronze Metallurgy

古代科技之光 中国 The Torch of Science and Technology in Ancient China

3
Bronze Linchpin
铜车辖

这件铜车辖为西周车马器。它是车轴上的销子，插入轴末端的方孔内，以防车轮脱出。

原件出土于河南省平顶山应国墓。

A kind of chariot implement dating from the Western Zhou Dynasty, it was used as a bolt which was inserted into the square hole at either end of the axle of the chariot in order to prevent the wheels from coming off.

The original was excavated from a tomb of the State of Ying located at Pingdingshan of Henan Province.

4
Bronze Guan
铜车輨

这件铜车輨为西周铜器。直径9.5厘米，是用来保护车毂的器件。

原件出土于河南省平顶山应国墓。

A bronze casting dating from the Western Zhou Dynasty, it was 9.5 centimeters in diameter and used as a protective device for the chariot hub. The original was excavated from a tomb of the State of Ying located at Pingdingshan of Henan Province.

5
Horse Yoke Ornaments
轭首、轭足

这组轭首、轭足为西周铜器。轭首、轭足是马轭的装饰物。原件出土于河南省三门峡虢国墓。

Dating from the Western Zhou Dynasty, the castings, literally called "yoke head" and "yoke hoof" in Chinese, were used as horse yoke ornaments. The original pieces were excavated from a tomb of the State of Guo located at Snamenxia, Henan Province.

92

这件铜轺车为东汉铜器。轺车为皇帝出行仪仗队的前导车之一，由车、马、伞、御奴组成。车为双曲辕，后部为长方形舆，轼中部插伞。御奴呈跪姿，双手拱举做执辔驾驭状。

原件出土于甘肃省武威市雷台汉墓。

Dating from the Eastern Han Dynasty, the chariot Yao was amongst the leading vehicles in the emperor's guards of honor during his majesty's inspection tour. It was composed of a vehicle, a horse, an umbrella and a horseman. This kind of light chariot had a hyperbolic shaft, with a rectangular sedan on the rear and an umbrella inserted in its center. The charioteer was in a kneeling posture with his hands holding the bridle.

The original chariot was excavated from a Han tomb at Leitai of Wuwei city, Gansu Province.

这件铜辇车为东汉铜器。辇车为出行仪仗队伍中官吏家眷乘坐的车。原件出土于甘肃省武威市雷台汉墓。

Dating from the Eastern Han Dynasty, the carriage Nian was part of the procession of the guards of honor and used by the family members of officials who accompanied the emperor going out on an inspection tour.

The original was unearthed from a Han tomb at Leitai of Wuwei City, Gansu Province.

这件铜斧车为东汉铜器。斧车为出行仪仗队伍中的前导车,是出行仪仗中的重要组成部分,车上立斧以示权威。

原件出土于甘肃省武威市雷台汉墓。

Dating from the Eastern Han Dynasty, the axe cart was amongst the leading vehicles used by the guards of honor during the emperor's inspection tour and an important part of the imperial procession. An upright axe was installed in the cart as a symbol of power.

The original was unearthed from a Han tomb at Leitai of Wuwei City, Gansu Province.

这件铜持戟骑士俑为东汉铜器。它由武士、戟、马、鞍组成。原件出土于甘肃省武威市雷台汉墓。

The casting, composed of a warrior, a halberd, a horse and a saddle, dates from the Eastern Han Dynasty. The original was unearthed from a Han tomb at Leitai of Wuwei City, Gansu Province.

前 221 年秦始皇统一中国，成为至高无上的统治者。作为中国历史上第一个大一统王朝——秦王朝的开国皇帝，他不遗余力地彰显自己的权力和威望。秦一号铜车马便是其死后陪葬品之一，代表皇帝出行仪仗队的前导车。图中所示是按照原物 1∶2 比例的仿制品。秦一号铜车马出土于秦始皇陵西侧墓，充分显示了 2200 年前中国青铜铸造工艺的先进水平及高超的加工技术。

In 221 B.C. Emperor Qinshihuang unified China and became its supreme ruler. As the first emperor of a unified China known as the Qin Empire, he used all occasions to show his power and grandeur. After his death, the bronze horsed chariots representing the leading vehicles in the emperor's guards of honor were buried with him, among other things, as burial objects. The set shown here was reconstructed on the scale of 1∶2 according to the original chariot, which was excavated from the west side of Qinshihuang's mausoleum. The making of the bronze horsed chariot fully demonstrates the exquisite casting and metal processing technology developed by the Chinese people some 2,200 years ago.

铜车马伞杆随秦始皇兵马俑一号铜车马一起出土。伞骨上集合了暗锁、双合页和插销等装置，标志着秦代青铜铸造工艺的先进水平及中国古代高超的加工技术。

Unearthed together with the bronze horsed chariot, the umbrella rod was composed of a built-in lock, double hinge, bolt and other devices. Not only does it represent the advanced bronze casting technique in the Qin Dynasty but it also demonstrates the sophisticated processing skills of the ancient Chinese technicians.

青铜冶铸
Bronze Metallurgy

1 犀牛尊 *Rhinoceros-shaped Zun*

犀牛尊是古代容酒备斟之器，因形似犀牛而得名。此为前11世纪至前771年犀牛尊的复制品，原件出土于陕西省兴平。犀牛嘴的一侧伸出一根管状的"流"以便倒酒。一般多用复合范工艺铸造。

This Zun ritual wine vessel was cast in the shape of a rhinoceros. It is a replica of an original dating back to the 11th century B.C. to 771B.C., which was found at Xingping in Shaanxi Province. The wine would have been poured through the spout to the side of the mouth. Normally it was founded with the compound-molds method.

2 铜爵 *Bronze Wine Vessel Jue*

爵是温酒、盛酒的器皿。左图为商代铜爵的复制品，原件于1976年出土于河南省安阳。铜爵前有长流，后有尖尾，上部略显宽大，为保持平衡，设计成中部束腰，下面三个细高的尖足，整体造型优美而雅致。

This is a copy of a Shang Dynasty Jue which used for warming or holding wine. The original was excavated in 1976 at Anyang in Henan Province. With a long narrow spout and an elongated tail, the vessel's upper portion looks fairly large. To avoid creating imbalance, the designer shaped it with a slender waist and three long, thin legs, resulting in a very elegant and beautiful piece.

3

提梁卣
Ritual Vessel You

卣是古代盛酒的器皿，有盖和提梁。图中纹饰精美的器物为商代提梁卣的复制品。此种礼器盛行于商晚期和西周。

The You or Ewer was a wine vessel distinguished by having a lid and a loop handle. This elaborately decorated example is a reproduction of a specimen produced under the Shang Dynasty. Ritual vessels of this kind were most popular during the late years of the Shang Dynasty and the Western Zhou Dynasty.

4

折觥
Zhe Gong

左图所示折觥为西周盛酒器的复制品。原件于 1976 年出土于陕西省宝鸡市。

This is a copy of a drinking vessel dating back to the Western Zhou Dynasty. The original was excavated in 1976 in Baoji of Shaanxi Province.

5

龙虎尊
Ritual Vase Zun with Dragon and Tiger Motif

尊为盛酒的器皿。右图为 1957 年出土于安徽省阜南的商代龙虎尊的复制品。尊上饰有蟠龙和双虎食人的图案。

A Zun was a vessel used to hold wine. This is a replica of the Shang Dynasty original found in 1957 at Funan in Anhui Province. It is decorated with figures of horned dragons and pairs of tigers.

青铜冶铸
Bronze Metallurgy

何尊为古代盛酒器。这种酒器上口侈大，腹部微鼓，下有圈足。它是西周时一位何姓的贵族用作祭祀的尊。原物于 1964 年出土于陕西省宝鸡市，用复合范工艺铸造。

An ancient wine vessel with flaring lip, swelling body and rimmed foot, He's Zun was ordered to be made as a sacrificial vessel by Duke He, an aristocrat living in the Western Zhou Dynasty. The original, cast with the compound-molds method, was excavated in Baoji of Shaanxi Province in 1964.

青铜钫为古代方口方身壶类盛酒器。

Fang belongs to the pot-type of wine vessels with a square mouth and round belly.

方座簋为古代盛食物或粮食的器皿，它圆形侈口，双耳或四耳。始于商中晚期，盛于周，衰于战国时期。

Gui is a round-mouthed bronze food vessel with two or four loop handles. It was used in ancient times for holding food or grains. Such food vessels first appeared in the middle and late period of the Shang Dynasty and became popular during the Zhou Dynasty. The Warring States Period saw their gradual decline.

乐器 *Musical instruments*

铜鼓为中国西南少数民族常用的乐器，是象征权力和财富的重器。原件为春秋战国时期铜器。

Symbolic of power and wealth, a bronze drum of this kind was popularly used as a musical instrument by the ethnic minority tribes living in Southwestern China. The original bronze drum dates from the Spring and Autumn and Warring States Period.

2 编钟 Chime-Bells

　　多个频率不同的钟按大小依次排列，构成合律合奏的音阶，称为编钟。它是古时祭祀或宴飨时用的乐器。编钟悬挂在木架上，有 3 枚一组的，还有多达 12 枚一组的。

　　演奏时，上层的编钟用来定调，下层的用来伴奏。测试表明，只要准确地敲击钟上标音的位置，每只钟都可以发出两个不同的乐音。

　　下图中为战国时期（前 475—前 221）编钟的复制品。其音质优美，音阶准确，至今仍可演奏乐曲。

Bells of different frequencies that were arranged according to their sizes and grouped together to form a harmonious scale are called "chime-bells". In the ancient times, they were often used as musical instrument in sacrificial rituals and banquets. Hung on wooden beams, each set of chime-bells could be composed of three or even a dozen bells.

During performance, the upper row of chimes was used to set the tune while those in the bottom row served to play the accompaniment. Tests have shown that if struck accurately on the correct inscribed position, the chimes would each produce two different tones.

The set shown here was reconstructed based on the chime-bells dating from the Warring States period (475B.C.–221B.C.). With fine tone quality and accurate scale, they can still be used to play music today.

生活器具 *Life utensils*

1

鉴是古代盛水或装冰的器皿，有架冰降温、盛水照容之用，盛行于春秋战国时期。图中所示为复制品，原物出土于安徽省寿县蔡侯墓。吴王光嫁女于蔡侯并铸此鉴作为陪嫁物，鉴内壁铭文有所记载。

A Jian was a tub-shaped vessel used for holding water or ice. It was in popular use during the Spring and Autumn and Warring States periods. When filled with water, it could be used as a mirror; when filled with ice, it could help bring down the room temperatures. On display here is an exact copy of a vessel found in the tomb of Marquis Cai in Shouxian County, Anhui Province. There is an inscription on the interior which tells how King Wu ordered the vessel to be made as part of the dowry for his daughter, who married Marquis Cai.

2

右图所示为辽宁省发现的11—13世纪龙洗盆的复制品。盆内铸有龙纹图案，标志着是特为皇宫制造的。当用手摩擦盆边的双耳时，手法得当，会产生共振的效果，盆中的水会如喷泉般溅起水柱。

Copy of a basin dating back to the 11th–13th centuries found in Liaoning Province. The dragon (Long) pattern on the interior suggests that it was specially made for the Imperial Court. If the handles are rubbed in the right way, water in the basin will spray like a fountain because of vibration.

鱼洗是从龙洗演变而来的，因盆内有鱼纹而得名。

A later variety of the Longxi, the Yuxi gets its name from the fish (Yu) motif on the interior.

康侯铜斧为西周铜器。铜斧上有"康侯"铭文。康侯是周武王的弟弟姬封，也是卫国始封之君。

原件出土于河南省浚县。

A casting dating from the Western Zhou Dynasty, the bronze axe bears the inscription "Marquis Kang", or Ji Feng, the younger brother of King Zhou Wu, who ruled as the first king of the State of Wei.

The original was unearthed from Junxian County of Henan Province.

　　下图为东汉（25—220）晚期的铜奔马复制品，它有一个诗意的名字"马踏飞燕"。原件于1969年出土于甘肃省武威市雷台。它造型构思巧妙，富有浪漫主义气息，而且精确地运用了力学的平衡原理。马的右后蹄踏着一只凌空飞翔的燕子，表现出天马行空的风驰电掣状。马踏飞燕反映了东汉时期杰出的制作工艺和铸造水平。

Dating back to the late part of the Eastern Han Dynasty (A.D. 25–A.D.220), the bronze galloping horse has a poetic name—"the galloping horse trampling on a flying swallow". The original was unearthed in 1969 at Leitai of Wuwei City, Gansu Province. The horse is smartly shaped and in a romantic mood. The flying swallow under its right hoof sets out that the horse is as swift as the wind and quick as lightning. The casting of the galloping horse reflects not only the outstanding workmanship but also the superb casting techniques attained during the period of the Eastern Han Dynasty.

　　透光铜镜是一种特殊的青铜镜，它正面微凸，不仅能照人，当平行光束照射在并不透明的镜面时，镜面的反射投影像还能映现镜背的纹饰和铭文。这种特殊的效果产生自铸造铜镜时的冷却与抛光过程。

The "transparent" bronze mirror is a special kind of bronze mirror; its obverse side is slightly convex. What is special about this mirror is that it can not only reflect one's image, but also faintly show the designs and inscriptions on its back when the parallel light beam shines on the surface of the mirror. This effect resulted from the cooling and polishing processes when making the mirror.

唐二十八宿铜镜
Bronze Mirror with 28 Lunar Mansions

海兽葡萄镜
Marine Animal Grapes mirror

　　齐刀币是春秋战国时期中国部分地区使用的货币，其他还有铸成锄头或齿贝形状的货币。原件发现于山东省淄博市。

Coins shaped like this were used in parts of China during the Spring and Autumn Period and the Warring States Period. Other coins were shaped like miniature hoes or cowrie shells. The original of this knife-shaped coin was found at Zibo of Shandong Province.

　　长信宫灯出土于河北省满城西汉中山靖王刘胜正妻窦绾墓。其通体鎏金，形状为一跪坐的宫女手持宫灯，整体由头部、身躯、双臂、灯座、灯盘和灯罩六部分组成，各部均可拆卸。宫女身着广袖长袍，左手持灯座，右臂高举与灯顶部相通，形成烟道。灯罩由两片弧形板合拢而成，可转动，用以调节光照度和方向。

　　器身刻有铭文，记载了该灯的容量、重量及所属者。因灯上刻有"长信"字样，故名"长信宫灯"。

Dating from the Western Han Dynasty, the original was excavated from a tomb in Mancheng, Hebei Province whose owner was Dou Wan, wife of King Liu Sheng of the State of Zhongshan in the Western Han Dynasty. Entirely gilded and in the shape of a kneeling court maid holding a lamp, the casting consists of six parts, namely, the head, the torso and the arms of the maid, as well as the lampstand, lamp tray and lampshade, all of them can be taken apart. The figurine, dressed in a robe with wide sleeves, is shown to hold the lamp stand with her left arm while her right arm is connected to the top of the lamp to form a chimney. The lampshade is composed of two roof tile-shaped panels that can be turned freely to adjust the brightness and illumination direction.

On the casting are inscriptions telling the capacity, weight and owner of the lamp. Two Chinese characters "chang xin" (eternal fidelity) are clearly marked on the lamp, therefore, it is also referred to as " Lamp from the Chang Xin Palace".

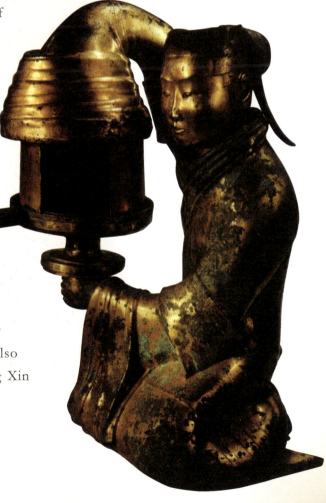

9 博山炉
Boshan Incense Burner

中国古人习惯将檀香木或其他香料放在熏香炉里焚烧，用来杀菌驱蚊或净化室内空气。我国熏炉种类繁多，其中以博山炉最为有名。这些熏炉造型雅致，炉盖镂空或炉身有孔洞。

The ancient Chinese had the tradition of putting sandalwood and other kinds of perfume inside incense burners in order to kill bacteria and mosquitoes, as well as to make the indoor air fresh. There was a great variety of incense burners, with the Boshan incense burners being the most famous; and they were all shaped elegantly. Some incense burners had hollowed-out lids; others had holes on their bodies.

兵器 *Weapons*

　　在中国如《武备志》一类的兵书中记载了各式中国传统兵器，包括剑、矛、戟及其他许多异形兵器。图中展示的是一些具代表性的兵器。

Chinese military manuals such as *Wu Bei Zhi* (*Treatise on Armament Technology*) record a variety of traditional Chinese weapons. They include swords, spears, halberds and a number of more exotic shapes. Shown here are some representative weapons.

1 人面钺
Tomahawk with Human
Face Motif

2 青铜戈
Bronze Dagger-axe

青铜冶铸
Bronze Metallurgy

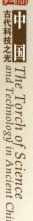

中国
古代科技之光
The Torch of Science and Technology in Ancient China

3
铜镞
Bronze Arrowhead

中国古代
科技之光

THE TORCH OF SCIENCE AND TECHNOLOGY IN ANCIENT CHINA

机械发明

Mechanical Inventions

科古技代

中国古代
科技之光

THE TORCH OF SCIENCE AND TECHNOLOGY IN ANCIENT CHINA

　　中国自古以农立国，传统农业和农业机械有着悠久的历史以及丰富的创造。从史前最简单的工具开始，至春秋战国时期（前770—前221）耕作机械的逐步形成、秦汉时期（前221—220）谷物加工机械的逐步发展、隋唐时期（581—907）农田灌溉机械的逐步完善，直到宋元时期（960—1368）各种农业机械的发展达到顶峰，农业机械在耕地、播种、收获、粮食加工、灌溉等方面发挥了巨大作用，促进了古代农业生产的发展。这些农业机械的种类繁多、构思巧妙、工艺精湛，有些沿用至今。

　　中国古代先民用自己的勤劳智慧，在衣食住行诸多方面都发明制造了许多生活所必需的工具和机械，这些实用性很强的生活工具，极大提高了人们的劳动效率和生活质量。

Ever since the ancient times, China has regarded agriculture as the foundation on which to build up the nation and make it stable. As such, it has a long history of traditional farming and agricultural machinery, as well as an unbelievable wealth of inventiveness in the mechanical field. Starting from the prehistoric period which witnessed the invention of the simplest tools, to the Spring and Autumn and Warring States periods (770B.C.–221B.C.) which saw the gradual evolution of farming machinery, to the Qin and Han Dynasties (221B.C.–A.D. 220) when the gradual development of grain processing machinery took place, to the Sui and Tang Dynasties (A.D.581–A.D.907) during which irrigational tools underwent gradual improvements, and all the way up to the Song and Yuan Dynasties (A.D.960–A.D.1368) when the development of all kinds of agricultural machines reached its peak. All these agriculture-related mechanical inventions had played a tremendous role in plowing, sowing, harvesting, grain processing, irrigation and so on, hence greatly promoting the development of agricultural production in ancient China. Thanks to their diverse types, smart designs and sophisticated workmanship, some of these agricultural mechanical devices are still being used today.

In terms of the basic necessities of life, the ancient Chinese, with their diligence and wisdom, had created numerous tools and instruments indispensable to their day-to-day lives. Because of their great practicality, these tools and equipment had greatly improved the work efficiency of the ancient Chinese people and the quality of their lives.

生产工具 *Production Tools*

西汉铜齿轮，原物出土于陕西省长安洪庆村（今西安市长安区）。

Shown here are the replicas of the bronze gears. Made during the Western Han Dynasty, the originals were excavated from Hongqing Village of Chang'an County (today's Chang'an District of Xi'an City) in Shaanxi Province.

汉代铁齿轮，原物出土于河南省鹤壁鹿楼村。

Shown here are the replicas of the iron ratchet wheels. Made during the Han Dynasty, the originals were unearthed from Lulou Village of Hebi City in Henan Province.

3 铁铧和鐴土
Iron Plow and Moldboard

西汉时期复合农具。原物出土于辽宁省辽阳市三道壕。铁铧为耕犁破土的锋刃，鐴土是耕犁的翻土器。鐴土与铁铧组合构成的复合装置，能将耕起的土垡破碎和翻转，更适于耕翻土地和开沟作垄。

This is the replica of a composite farm tool which first appeared in the Western Han Dynasty. Excavated from Sandaohao, Liaoyang City, Liaoning Province.
Iron plow is the sharp blade for unearthing, while moldboard is the unearthing tool on a plow. Combining the two kinds of tools as a composite device. It is more suitable for plowing and guttering.

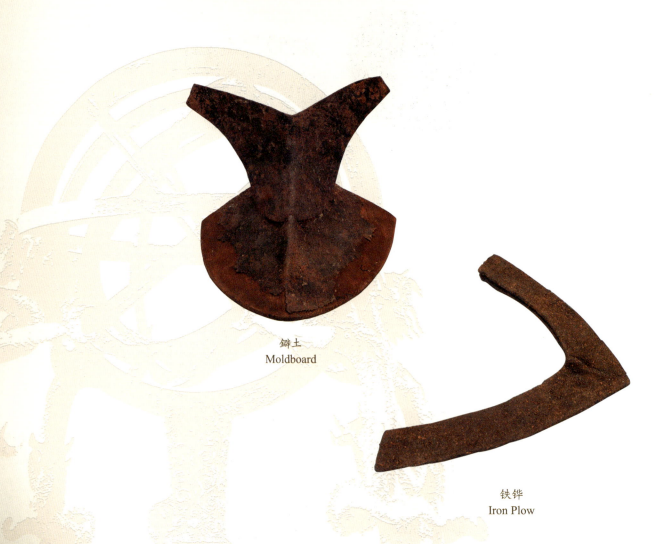

鐴土
Moldboard

铁铧
Iron Plow

　　耕犁是古代用于犁地的主要农具，有 5000 年的历史。我国早在新石器时期就出现了石犁。在战国时期，中国人利用先进的金属冶炼技术率先制造出了铁犁。到了汉代，由两头牛牵引的耕犁已经广泛地应用。在唐代，为了便于耕作稻田，江东犁（一种水田曲辕犁）应运而生。

Used as a major farm tool by Chinese farmers even since the ancient times, plough has a history of some 5,000 years. As early as in the Neolithic Age, there were stone ploughshares. Advances in metal production and in design enabled the Chinese to produce the first iron ploughshares during the Warring States Period. During the Han Dynasty ploughs drawn by two cows were already in popular use. In the Tang Dynasty, in order to facilitate the ploughing of paddy fields, the Jiangdong ploughs (paddy ploughs with curved shafts) came into being.

5 耧车
Seed Drill

　　耧车是开沟、播种、覆土三功一体的农具，能够极大地提高播种的速度和质量。它的发明标志着人类手工播种的巨大进步。

　　耧车有独脚耧、两脚耧和三脚耧之分，可以由人力或畜力驱动。自西汉时期发明沿用至今。

Invented during the Western Han Dynasty, the seed drill could accomplish the work of ditching, sowing and soil refilling at one time, thus greatly increasing the speed and quality of sowing. Its invention marked a great advance on the manual scattering of seeds.

Seed drills are divided into single-row seed drills, two-row seed drills and three-row seed drills. They could be drawn by farmers or by animals and are still in use in some areas of China today.

这组石磨盘、石磨棒是距今8700—6800年前的裴李岗遗址出土的新石器时代原始粮食加工工具复制品。

Shown here are the replicas of the grindstone and stone roller, the originals were excavated from the Peiligang site.
Dating back to around 8,700 years to 6,800 years ago, these are grain processing tools of the Neolithic Age.

7

石 磨

Stone Rotating Mill

石磨是出现在晋代（265—420）的粮食加工工具。它由人力或畜力驱动。将谷物倒入磨盘上的孔内，使其进入磨盘之间的缝隙，经碾磨后变为粉状。直至今日，这样的石磨仍在中国的部分地区使用着。

A grain processing tool invented in the Jin Dynasty (A.D.265-A.D.420), it could be driven either by human or by animal power. Grains were poured in through a hole in the upper stone and the resulting flour spilled out through the gap between the two millstones. Mills like this are still in use in some areas of China today.

8 扇车
Winnowing Machine

扇车是汉代出现的清除糠秕的农机具。它由支架、外罩、风扇、进料斗和调节器等部件组成。操作时打开调节器转动风扇摇柄，使谷物从料斗缓慢下降，杂质被吹走。这种脱壳工具沿用至今。

A farm machine used for husking grain, the winnowing machine first appeared in China in the Han Dynasty and still in use in some areas of China today. It consisted of the stand, the outer covering, the fan, the hopper and the regulator etc. The operator turned the fan in order to open the regulator so that the grain could fall down slowly from the hopper and the impurities be blown away.

下图所示的屯溪磨坊是一个综合性粮食加工厂的模型，出现于明代安徽省皖南屯溪地区。它利用水轮能同时驱动连击水碓和磨盘对谷物进行脱壳和面粉制作。它的发明体现了 14 世纪的中国在水力机械方面的成就。

Shown here is model of a comprehensive grain processing mill typical of those used at Tunxi, Anhui Province during the Ming Dynasty. Its waterwheel could simultaneously drive both a series of trip hammers to husk rice and sets of millstones to produce flour, which is an indication of the further advances made in utilizing water power by the 14th century.

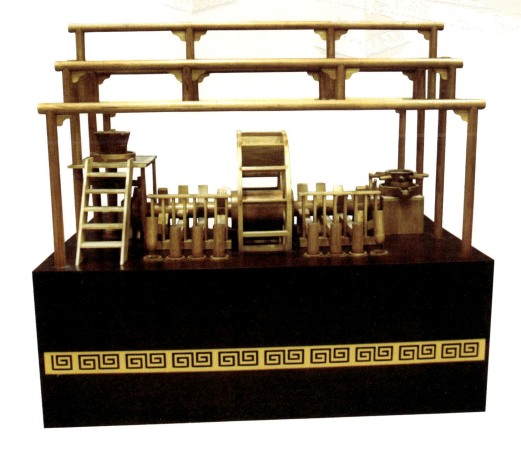

10

桔槔
Jie Gao (Shaduf)

桔槔是春秋时期（前 770—前 476）出现的利用杠杆原理制成的井上汲水工具。古人在树立的架子上加上一根木杆，中间做支点，一端悬挂重物，一端悬挂水桶。将水桶放入水中打满水后，由于杠杆另一端的重力作用，便能轻易把装满水的水桶提拉上来。

In Spring and Autumn Period (770B.C.–476B.C.), a new human-powered irrigation tool—Jie Gao was invented, It adopts the lever principle to leverage water. Jie Gao actually is to add a wooden stick on a shelf. The fulcrum is in the middle, with one side for hanging a weight, the other side hanging the bucket. When the bucket is full of water, it can be lifted very easily with the weight of the other side of the leverage.

11

辘轳
Well Windlass

辘轳是一种汲水工具，春秋战国已有记载，宋代时出现曲柄辘轳并沿用至今。这种工具通常安装在井口中央，通过辘轳汲水非常方便并且能够节约劳动力。

As a water drawing tool, windlass was first recorded in the Spring and Autumn and Warring States periods. In the Song Dynasty, the crank windlass appeared and are still in use today. It was installed in the middle of the wellhead, and by means of which the drawing of water from the well was made convenient and less laborious.

翻车又名"龙骨水车"，是中国最古老的农业灌溉机械之一。由东汉时期（25—220）的毕岚发明，并在三国时期（220—280）由马钧加以改进。

翻车由传动机构和水槽构成。传动机构是由链轮、链条紧贴水槽内壁直立的木板组成的。它的链条（龙骨）与链轮（大轮、小轮）啮合。以链传动来实现运动的传递，做到连续提水输送。它可以将流水连贯顺畅地引入农田中灌溉作物。翻车分手摇翻车、脚踏翻车，也可利用畜力、风力驱动。

One of the most ancient forms of agricultural irrigation equipment in China, the chain pump, also known as the "dragon's backbone water lift", was created by Bi Lan of the Eastern Han Dynasty (A.D.25–A.D.220) and further improved by Ma Jun during the Three Kingdoms Period (A.D.220–A.D.280).

It consisted of an inclined wooden trough in which moved a series of connected wooden boards, standing upright and closely fitting the trough. The chain (dragon's backbone) and the chain wheels (a large wheel at the top of the trough and a smaller one at the bottom) mesh. The whole system resembled a continuous circular belt running over the chain wheels, raising water from the river to the paddy fields. Many different versions of the chain pump were built and they could be operated by hand, feet, animals or water power.

筒车是利用水力自动提水的装置，又称为"天车"。最早记载出现在 7 世纪唐代早期。它是由一个竖立的大水轮和边缘安装的竹制或木制的小桶组成的。水轮下端浸入河流，通过水流驱动大立轮旋转，小桶灌水后随立轮上升至顶部，将水注入水槽流向灌溉渠。

An automatic pump powered by water, the bucket wheel, as known as "the sky wheel", came into use in the 7th century during the early period of the Tang Dynasty. It was a large, vertical waterwheel with bamboo or wooden buckets attached to its rim. The lower edge of the wheel dipped into the river and was turned by the water current. The buckets were thus filled, carried upward and discharged into a flume running into an irrigation ditch. This device may also be driven by animal power.

舟磨是一种利用水流运作的粮食加工机械。在 13 世纪元代王祯的《农书》中有简要介绍。

A grain processing machine powered by water, it was briefly recorded in the 13th century *Book of Agriculture* by Wang Zhen of the Yuan Dynasty.

磨车是一种行进式粮食加工机械，又名行军磨，出现于南北朝时期。

A mobile grain processing machine, the mill carriage is also known as "field mill". It first appeared during the period of the Northern and Southern Dynasties.

16

舂车

The Husking Chariot

舂车通常将碓臼置于马车或其他畜力车上，是一种可以一边前进一边加工粮食的机械，最早出现于晋代。

A kind of mobile grain processing machine, the husking chariot was in fact, a pestle and mortar mounted on a horse-drawn chariot or other similar animal-powered vehicles. It made its first appearance in the Jin Dynasty.

17 刮车
Scoop Wheel

刮车是一种中国古代简便的手摇农田灌溉工具。

It was a kind of simple hand-operated irrigation tool used in ancient China.

18 井车
Barrel-Chain Pump

井车是从深井中提水进行灌溉的工具，大约产生于唐代。

Coming into use around the Tang Dynasty, it was a device used by the ancient Chinese for lifting water from a deep well to irrigate paddy fields.

水排是在汉代建武七年（31）由杜诗发明的一种水力机械装置。在铸造过程中，这种水排通过水力驱动的杠杆装置推拉皮囊向炼铁炉内送风，提高炉温。水排中集成了轮轴、凸轮轴以及杠杆的原理。

Invented and made by Du Shi in the 7th year (A.D. 31) of the reign of Jianwu in the Eastern Han Dynasty, hydraulic bellows like this were used in foundries for blowing air into the iron smelting furnaces by means of leather bags worked by waterwheel-driven levers. This mechanism made comprehensive use of wheel axle, camshaft and levers etc.

20 立式风车
Vertical Windmill

古代科技之光 中国 *The Torch of Science and Technology in Ancient China*

除利用水力外，中国古人也使用风力来驱动机械装置。立式风车发明于宋代（960—1279），它能够带动龙骨水车等机具以提取海水制盐或提水灌溉农田。

我国古代的立式风车与欧洲风车的不同之处在于，其扇面是采用垂直设置的方式，任何方向的来风都能驱动，使其能不停地工作。

The ancient Chinese harnessed not only water power but also wind power. Invented in the Southern Song Dynasty (A.D.960–A.D.1279), windmills like this were used extensively either to lift seawater for salt-making or to drive such irrigational devices as the "dragon's backbone water lift" to irrigate paddy fields.

In contrast with their European version, the Chinese windmills were set up vertically, hence guaranteeing constant operation regardless of the direction of the wind.

1

铜卡尺
Copper Caliper

　　铜卡尺是西汉末年王莽建立新朝后（9—23）创制的量具。用于测量器物的直径和深度。

A kind of measurement tool created during the closing years (A.D.9–A.D.23) of the Western Han Dynasty when Wang Mang founded the pseudo Xin Dynasty, it was used for measuring the diameter and depth of an implement.

2

天平和环权
Scales and Ringed Weights

　　天平和环权为战国时期制品，是利用杠杆原理制造的衡器。

Invented during the Warring States Period, the scales and ringed weights was a kind of weighing apparatus made based on the lever principle. The ring-shaped balancing weights were made of bronze and used on the scales.

机械发明
Mechanical Inventions

3 铜合页
Bronze Hinges

铜合页为春秋战国时期制品，原物出土于河南省辉县琉璃阁甲墓。

Shown here are the replicas of the bronze hinges. Dating from the late part of the Spring and Autumn Period and the Warring States Period, the originals were excavated from Tomb Group A of the Liulige (literally, Pavilion with Glazed Roof Tiles) ruins in Huixian County of Henan Province.

4 尖底瓶
Bottom-Pointed Pitcher

尖底瓶是古代取水工具，当瓶底接触水面后，由于重力作用，瓶子自然倒向一侧，水自瓶口流入瓶内，达到一定水量时，瓶体自动立起，瓶身垂直下沉，从而取满水而滴水不漏。

It is a kind of water fetching tool used in the ancient times. When used and due to the gravity factor, the pitcher would naturally lean to one side as soon as its bottom hit the surface of the water. Then, water would flow into the pitcher from its mouth. When the pitcher was filled with a certain amount of water, it would automatically resume its upright position and sink gradually in a vertical manner. After being fully filled with water, the pitcher could be lifted up from the water to complete the water fetching process.

攲器是由生活用具演变成的具有象征意义的警戒之物。攲器在未装水时器口前倾、装少量水时器身直立，注满水时器物倾覆水倒净，器物自动复原。攲器试图向人们传达的信息是：满招损，谦受益。

Evolving from a daily-used utensil, it was a vessel with symbolic warning significance. When empty, the suspended cup-shaped vessel was slightly inclined; when filled with a small amount of water, it would stay upright. However, when fully filled, the vessel would turn over, emptying all the water. Then it would resume its original position.

The lopsided vessel was designed to convey the message: haughtiness invites disaster; humility reaps benefits.

银香薰发明于西汉时期，是一种用在被子或毯子里的薰香器具，亦称"被中香炉"。其原理与现代陀螺仪中的万向支架类似。银香薰外壳镂空，内有 2 个相互垂直、自由转动的圆环，安置在内环中的半球形香炉则通过 3 个相互垂直的轴旋转，无论球形的香薰如何旋转，位于中心的香炉始终保持水平状态。

Invented in the Western Han Dynasty for scenting quilts or blankets, also known as "censer in quilts", such censers made use of principles similar to that of a universal support used in a modern gyroscope. Inside the hollowed-out sphere of the censer were two rings vertical to each other and could be turned freely. The body of the censer was supported inside the inner ring and could rotate around three axis vertical to each other. No matter how the spherical censer rolled, its hemispheric body in the central position would remain horizontal.

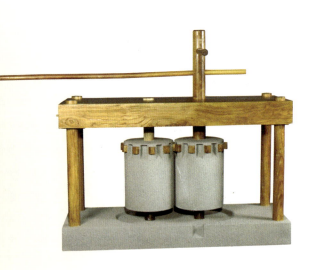

古代科技之光
中国
The Torch of Science and Technology in Ancient China

7
榨糖机
Sugar Squeezer

榨糖机最初记载于《天工开物》一书中，发明于17世纪的明代。将甘蔗作为原料放入两个紧密啮合并旋转的石辊中碾过，甘蔗的汁液被榨出，并沿导流槽汇集在一起，随后被熬制成黄糖。

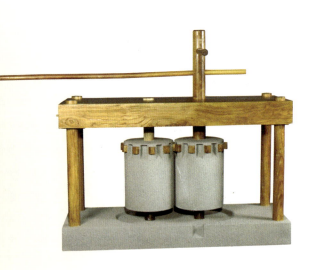

Sugarcane as the raw material for sugar-making was fed into the gap between the two closely joggled rollers for extraction of its juice, which would flow out along the guiding gutter. The squeezed-out juice would then be gathered for making brown sugar. Records of such sugar-making method can be found in the 17th century Ming Dynasty book *Tiangong Kaiwu* (*The Exploits of Nature's Works*).

8
榨油机
Oil Press

压榨法取油在我国有着悠久的历史。在宋、元、明、清近千年的历史岁月里，榨油机为不断增长的城市居民提供了充足的食用油。

Pressing is a kind of oil-making method with a longstanding history in China. During a long period of nearly one thousand years spanning the Song, Yuan, Ming and Qing Dynasties, it helped ensure the ample supply of edible oil for residents of the ever-growing cities in the country.

中国古人发明了种类繁多的锁具，如簧片锁、木销锁、密码锁以及定向锁等。这些锁具在人们的日常生活中发挥了不可或缺的作用。

The ancient Chinese invented a relatively great variety of locks, including but not limited to spring lock, wooden bolt lock, password lock and directional lock etc. They all played an indispensable role in people's daily lives.

10

竹蜻蜓

Bamboo Dragonfly

竹蜻蜓是一种中国传统玩具，双手轻轻搓动竖杆后放开，它就能旋转着飞上天空。西方传教士将竹蜻蜓称为"中国螺旋桨"。

A traditional Chinese toy, it could fly up into the sky in a spiral if the vertical shaft was slightly rubbed with hands and then let go. For this reason, the bamboo dragonfly was once called "the Chinese spiral" by the Western missionaries.

11

地动仪为最早的验震装置。古代科学家张衡于东汉阳嘉元年（132）在国都洛阳制成。当发生地震时，地动仪内部的都柱会向震中方向倾倒，触动机关，使龙嘴口中的铜珠掉落在蟾蜍口内。通过都柱倒向的位置和铜珠掉落的响声，可以判断出震中的方位。

The first of its kind in the world, this instrument was designed by Zhang Heng and built in the national capital of Luoyang in 132, the 1st year of the reign of Yangjia during the Eastern Han Dynasty. When an earthquake occurred, the tremor tilted the Duzhu inside the seismograph, causing it to press on the jaws of the dragon facing the epicenter and to drop the bronze ball from its mouth into that of the toad sitting below. The clanging of the Duzhu and the fall of the ball would thus sense the occurrence of the earthquake and also point the direction of its source.

12

绿釉陶磨坊是汉代的一种陪葬品，原物出土于河南省灵宝市张湾乡。从正面可以清晰地看到舂捣装置和石磨。

A Han Dynasty burial object, the original was unearthed from a tomb located at Zhangwan of Lingbao City in Henan Province. From the open side of the mill model, a husking device and a millstone can be seen clearly.

13

陶风车和陶米碓是汉代的一种陪葬品，原物出土于河南省济源泗涧沟的汉墓。

The original was excavated from a Han tomb located at Sijiangou in Jiyuan City of Henan Province, this burial object dates far back to the Han Dynasty.

军事应用 _Military Applications_

1

铜镞
Bronze Arrowhead

铜镞是一种商代兵器。

A kind of weapon dating from the Shang Dynasty.

2

铜弩机
Bronze Trigger

铜弩机是一种古代兵器，原物为三国时期魏正始二年（241）造。

The original of this bronze trigger was made in 241, the 2nd year of the reign of Zhengshi in the State of Wei during the Three Kingdoms Period.

3

强弩
Crossbow

弩出现于战国时期，到汉代时已普遍使用。

Crossbows first appeared in the Warring States Period, but their widespread use did not occur until the Han Dynasty.

The Torch of Science and Technology in Ancient China
古代科技之光

中国

134

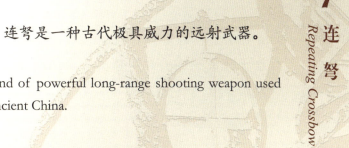

连弩是一种古代极具威力的远射武器。

A kind of powerful long-range shooting weapon used in ancient China.

三弓床弩是中国古代大型远射兵器。

A kind of heavy long-range shooting weapon used in ancient China.

机械发明
Mechanical Inventions

6 巢车
Nest Cart

巢车是一种中国古代的装甲侦察车，它出现于春秋战国时期，通过绞盘将像鸟巢一样的装甲侦察平台提升到高处，侦察员则通过瞭望孔观察整个战场及敌军布阵。

Used in ancient China as an observation post for military purpose, the nest cart first appeared in the Spring and Autumn and Warring States Period. The nest-like armored spotting platform could be raised with a windlass to give the observer a clear view of the battlefield and the enemies' deployment.

1

独轮车
Wheelbarrow

　　独轮车是一种低重心的轻型交通工具，发明于汉代。独轮车通常用于运送货物，偶尔载人。这种工具能够长距离运送货物，穿越狭窄崎岖的山村小路。

　　独轮车的设计特点是轮子几乎置于运载物品的货架中间，货物的重量几乎全部由车轮承载，这种巧妙的设计减轻了操作者的劳动强度。

As a light transportation means with low center of gravity, wheelbarrows first appeared in the Han Dynasty. They were used mainly for transporting loads, and occasionally for transporting passengers, over long distances, along narrow paths and over rough country.

The design of wheelbarrow placed the wheel almost under the center of the load. As a result, the weight was largely carried by the wheel and there was very little load on the handles, making the transportation relatively less laborious for the operator.

机械发明
Mechanical Inventions

2 记里鼓车
Odometer

　　记里鼓车为古代计里程的车辆，出现于西汉。它主要用作古代帝王出行时的仪仗车辆。该车使用减速齿轮系驱动凸轮轴。凸轮轴每旋转一周，车顶上的木头人敲击一次，宣布已经行驶了一里（0.5千米）。

Used for marking the distance travelled, the odometer cart was invented in the Western Han Dynasty. It was mainly used for ceremonial purpose when the emperor went on a journey. The axle of the cart was fitted with a reduction gear train which drove a camshaft. Each revolution of the camshaft would alternately activate a wooden figure on top of the cart to strike the drum once, announcing one li (half a kilometer) had been covered.

　　指南车为古代帝王出行时仪仗车队中指示方向的车辆，由三国时期的马钧发明。令车顶的小人指向南方，并以车轮、立轴与各种齿轮的复合运动为基础，不管车辆向任何方向行驶，小人所指示的方向始终为南方。

A direction indicating chariot chiefly used for ceremonial purpose when the ruler went on a journey, it was invented by Ma Jun of the Three Kingdoms Period. Set to point south, the figure on top of the pole was connected to the wheels via a differential gear, so that it would remain pointing south no matter in which direction the chariot was traveling.

中国古代
科技之光

THE TORCH OF SCIENCE AND TECHNOLOGY IN ANCIENT CHINA

Textile and Embroidery

中国是世界上最早养蚕并进行丝织的国家。早在六七千年前，人们就懂得用麻、葛纤维为原料进行纺织。公元前 16 世纪（殷商时期），产生了织花工艺和辫子股绣。公元前 2 世纪（西汉）以后，随着缫车、纺车、斜织机和提花机的发明，我国纺织、刺绣技术迅速提高。不但能织出薄如蝉翼的罗纱、构图千变万化的锦缎，还能绣出巧夺天工、琳琅满目的绣品，使中国在世界上享有"东方丝国"之称。从汉至唐，中国的丝织品经丝绸之路传入中亚、西亚和欧洲大陆，对世界文明产生了相当深远的影响。与此同时，丝绸之路也成为东西方交流的重要通道，也是今日"一带一路"倡议（丝绸之路经济带与 21 世纪海上丝绸之路）的本源。

Sericulture and silk weaving were first originated in China. 6,000–7,000 years ago. Chinese already knew how to weave with fibers of hemp and use kudzu vine as raw materials. Pattern weaving and "braid embroidery" emerged in China in the 16th century B.C. (the Shang Dynasty). After the 2nd century B.C. (the Western Han Dynasty), with the invention of silk reeling machine, spinning wheel, slant treadle loom and jacquard loom etc., the techniques of weaving and embroidering rapidly improved, and making it possible for the Chinese to produce gauze and leno as thin as a cicada's wings, as well as brocades of colourful designs. China hence enjoyed the reputation of "the Oriental Country of Silk" in the world. During a long period from the Han Dynasty to the Tang Dynasty, the Chinese silk products were traded to the Middle and West Asia and Europe via the famous Silk Road, having a far-reaching impact on the world civilization and constituting an important part of the world heritage of science and culture. Meanwhile, the Silk Road itself was later to take on great significance for East-West contacts; it also serves nowadays as the origin of today's Belt and Road Initiative (the Silk Road Economic Belt and the 21st Century Maritime Silk Road).

1 骨针 *Bone Needle*

骨针出现于旧石器时代晚期（距今约1万年），该物原件出土于北京市房山区周口店。

Dating back to the late part of the Paleolithic Age (about 10,000 years ago), the original needle was excavated at Zhoukoudian of Fangshan district, Beijing.

2 骨梭 *Bone Shuttle*

骨梭原件出土于山东省泰安市大汶口，距今6400年。梭为织布时牵引纬线的工具。

Dating back to the Neolithic Age (about 6,400 years ago), the bone shuttle was used as a weft traction tool. The original was excavated at Dawenkou of Tai'an City in Shandong Province.

3 纺轮 *The Spindle Wheel*

纺轮是在新石器时期就已出现的最原始的纺织工具，有石质、骨质、陶制和玉质等，形状有圆形、球形、锥形和齿轮形等。

As one of the most primitive mechanized textile tools invented in conjunction with silk production, the spindle wheel first appeared in the Neolithic Age and could be made of different materials such as stone, bone, clay and jade etc. Normally, the early Chinese spindle wheels were in the round, conical, spherical and gear shape.

纺织刺绣
Textile and Embroidery

143

纺坠由纺轮和捻杆组成。纺轮转动时，由自身的重力作用使纤维直径变小且长度增大，旋转产生扭矩将受拉的纤维（麻或纱）捻成麻花状，使纤维捻合或续接。

The spindle whorl consists of the spindle wheel and twisting rod. Using its own gravity, the spindle wheel rotates to make the fiber thinner and longer. The rotation produces a torque, forcing the twisted fibers (hemp or yarn) down in spirals to adhere to each other and form a continuous thread.

右图所示的铜片于 1950 年出土于河南省安阳大司空村殷墟，距今 3000 多年。铜片上有细绢遗痕，因为这件铜片当时是用细绢包裹的，绢腐烂后留下了痕迹。它反映出当时缫丝和织绸技术都已达到很高水平。

Vestiges of fine silk can be seen on this bronze fragment. The silk, which was used to wrap up the bronze ware, decayed over the years and hence left its vestiges on the bronze fragment. It is an indication of the high level of silk reeling and weaving technology attained at that time.

Excavated from the Shang Dynasty's ruins in Anyang of Henan Province in 1950, the original bronze fragment was made more than 3000 years ago.

青铜贮贝器的盖子，原物出土于云南晋宁石寨山遗址，距今 2000 多年，它的出土证实了中国早期就出现了纺织技术。盖面铸有 18 人，鸡犬各 1 只，案 1 张。各人的服饰、发式都不同。较大的女奴隶主坐于案前，监督奴隶进行纺织。奴隶们有用踞织机织麻布的；有将织成的布匹上光的；有理麻的；有检验的。纺织贮贝器盖形象地反映了 2000 多年前古滇族人从事纺织的劳动场面，亦是古代称为"手经指挂"的织造实况。

Dating back at least 2,000 years, the original was unearthed from the Shizaishan ruins in Jinning of Yunnan Province. Evidence of the early existence of Chinese weaving technology is shown on the cover of an ancient bronze vessel which used for storing shell coins. It has been applied with a group of 18 figures cast in the form of seated female slave weavers, as well as a chicken, a dog and a desk. With different dresses and hairdos, the women are working with a simple body-tension loom in which the warp threads are stretched between two beams, one of which is attached to the weaver's body while the other is secured by her feet. The centrally seated woman, a slave owner, looms large and is supervising the work of the slaves. The cover gives a vivid reflection of the working scene of the ancient people in Yunnan some two thousand years ago.

纺织刺绣
Textile and Embroidery

7 手摇纺车
Hand-operated Spinning Wheel

手摇纺车是一种纺纱工具，出现于汉代。它主要由一个大绳轮和一个小圆锭组成。大绳轮通过皮带或绳索带动小圆锭快速旋转，实现了麻或纱的自动加捻。

A kind of yarn spinning tool which first appeared in the Han Dynasty, this hand-operated spinning wheel mainly consists of a large rope sheave and a small round spindle. The former drives the latter to rotate rapidly via a belt or a rope, making the automatic twisting of hemp or yarn technically possible.

8 脚踏纺车
Treadle Spinning Wheel

脚踏纺车出现于宋代，由纺纱机构和脚踏机构组成。脚踏机构通过曲柄带动纺纱机构（绳轮和圆锭）转动，完成加捻牵伸工作。这种纺车通过脚踏机构提高了绳轮的牵引力，可双手操纵纱线，控制更多的圆锭，从而提高了生产效率。

First appeared in the Song Dynasty, the treadle spinning wheel consists of the yarn spinning mechanism and the treadle mechanism. Through a crank, the treadle mechanism activates the yarn spinning mechanism (the rope sheave and the round spindles) to rotate in order to accomplish the work of twisting and traction. Such kind of spinning wheel, via its treadle mechanism, greatly enhances the traction force of the rope sheave, making it possible for the weaver to use both hands to manipulate the yarns and control more round spindles, hence increasing productivity.

下图所示为北宋（960－1127）王居正画的《纺车图》。

Shown here is the painting of a spinning wheel by Wang Juzheng of the Northern Song Dynasty (A.D.960－A.D.1127).

10 脚踏缫丝车 *Treadle Silk-reeling Wheel*

　　脚踏缫丝车出现于宋代，在手摇缫丝车的基础上发展而成。脚踏缫丝车在丝框的曲柄处接上连杆并和脚踏杆相连接，用脚踩踏杆做往复摆动，通过连杆带动丝框曲柄，利用丝框回转时的惯性，使其保持连续转动，带动整台缫丝车运转。这样就可由一人完成索绪、添绪和转动丝框的工作，提高效率。

Treadle silk-reeling wheel, developed on the basis of its earlier hand-operated version, was invented in the Song Dynasty. It has slotted rod at the crank of the reel to connect to the footrest lever; reciprocating motion can be realized after stamping the treadle constantly, the slotted rod will drive the crank of the reel; the reel rotating inertial will then keep the continuous rotation and drive the whole reeling wheel. The work of groping, attaching and wireframe rotating can be effectively done by a single reeler, therefore resulting in increased productivity.

11

水转大纺车出现于南宋时期（1127—1279），元代（1206—1368）盛行于中原地区（黄河中下游地区），用于加工麻纱。

它是一种相当完备的机器，与近代纺纱机的构造原理基本一致。水转大纺车已具备动力机、传动机构和工具机，是当时世界上最先进的纺纱机械。

Used for spinning ramie into yarn, this water-driven spinning wheel was a major achievement in mechanical engineering in ancient China. It first appeared in the Southern Song Dynasty (A.D.1127–A.D.1279) and later became prevalent in the Central Plain Region (comprising the middle and lower reaches of the Yellow River during the Yuan Dynasty (A.D.1206–A.D.1368).

As a rather sophisticated spinning machine, the water-driven large spinning wheel, with its motor, transmission mechanism and tools, had basically the same structural principle of a modern spinning machine. It was the most advanced spinning machine in the world at that time.

　　斜织机是一种中轴式踏板织机，适于织造平纹素织物，因织机的经面与水平机座呈 50～60 度的倾角，故称斜织机。它是伴随着原始织机演变而成的具有牵伸、开口、打纬等机械功能的织机，在我国汉代已普遍推广。

First appeared in the Han Dynasty, the slant treadle loom is a kind of center shaft-treadle looms. The loom was so named because its warp thread panel is at a leaning angle of 50° ~ 60° against its horizontal base, which is suitable for weaving plain weave fabric. It evolved from the original weaving machines and has such mechanical functions as drafting, opening and beating-up. As early as in the Han Dynasty, such slant treadle loom was already in widespread use.

13
立织机
Vertical Treadle Loom

立织机最早的形象出现于甘肃省敦煌五代时期内壁画《华严经变》图中。踏板立机原理与踏板卧机相同，只是经线的方向垂直于地面。提经、投梭、打纬都在立架提挂的部件联动中完成，因此称为立织机。立织机最突出的特点是由两片踏板通过中轴的转动来控制单片综的升降而形成梭口，织机的自然梭口由一组分经机构开启。织机在运动中经轴也随踏板升降，使经丝的张力基本保持不变。立织机是中国古代踏板织机中最巧妙、最出色的一种。

First appeared in the Five Dynasties period, the vertical treadle loom has the same working principles as the horizontal treadle loom, except that in the vertical treadle loom the direction of the warp threads is vertical of the ground and the weaving processes, such as warp lifting, weft shuttle throwing and beating-up are done through the chain reaction of the components fixed on an upright framework, hence its name. The most prominent feature of the vertical treadle loom is that its two treadles control the ups and downs of the single heddle by working the rotation of the shaft, thus forming an opening for the shuttle. The natural opening for the shuttle is opened through a warp dividing mechanism. As a result, in the weaving process the warp shaft is raised or lowered along with the treadles, so that the tension of the warp threads basically remains unchanged. Of all the treadle looms developed in ancient China, the vertical treadle loom was the smartest and most outstanding one.

多综多蹑织机是一种综片式提花织机，出现于汉代。其原理是将图案信息按其中所含纬线的数量分成若干份，每一份的图案信息储存在一片综片内，若干个综片储存一个完整的图案。织造时，一蹑控制一综，每提升一次综片，预先穿入综片的不同色彩经线压在一根纬线上，织物得到一纬图案。综片依次提升，纬纬相加在织物上组合成完整图案。

A heddle-type of jacquard weaving loom, the multi-heddle and multi-treadle loom first appeared in the Han Dynasty. Its working principle is to divide the pattern information, based on the number of weft threads contained, into numerous pieces, and then every piece of pattern information is stored inside a single heddle. The myriad heddles combine to store the complete information of a full pattern. In the weaving process, one treadle controls one heddle; each time the heddles are lifted, the warp threads of different colors threaded through the heddles in advance would press on a single weft thread, forming a weft pattern on the fabric. When the heddles are lifted one by one in given order, the weft threads combine to form a complete pattern on the fabric.

15

小花楼织机出现于初唐并兴盛于唐以后各代。它机身平直，中间耸起花楼。楼上悬挂储存提花信息的花本，又称束综提花机。织造时由两人操作，一人坐于花楼一侧，按花本顺序拉动束综，提升相应起花部分的经线，以使经线形成开口，由另一人进行打纬织造。

First appeared in the early part of the Tang Dynasty and became prevalent in the ensuing dynasties after Tang, the small double drawloom has a level and straight body, as well as a vertical tower in its middle part. Hung from the top of the tower is the huaben, or pattern base, which contains the information of the jacquard weave. As such, it is sometimes referred to as the "bunched heddles jacquard loom". As its name suggests, weaving on the loom requires two weavers to work in partnership. One weaver is stationed above; and in predetermined sequence, pulls the bunched heddles to change the position of the corresponding warp threads to form openings. The other weaver sitting below controls beating-up and weaving weft by throwing the weft shuttle between the warps to form the raised pattern on the fabric.

　　大花楼提花机最早出现于唐中晚期并盛行于唐以后各代，是中国古代织造技术的最高成就。因它在机身的后部高出一个悬挂花本的提花装置，其形状似高楼，故被命名为"花楼织机"。它的提花的工艺方法源于原始腰机挑花及多综多蹑提花原理。提花的方法通常采用一蹑（脚踏板）控制一综（吊起经线的装置）与提花同时作用来完成织造花纹的任务。其最突出特点是能够织出复杂的、花形循环较大的花纹。云锦就是由大花楼提花机织造的。

First appeared in the middle and late part of the Tang Dynasty and became prevalent in the ensuing dynasties after Tang Dynasty, the large double drawloom marks the highest achievement of ancient China in terms of weaving technology. Uniquely equipped with a towering bunched heddles jacquard device that controls the patterning, the loom is also vividly called "weaving loom with a flower tower" in Chinese. The jacquard techniques of the large double drawloom originated from hand-stitching of the primitive body-tension loom (back-strap loom) and the jacquard principles of multi-heddle and multi-treadle loom, with each treadle (foot pedal) controlling one heddle (the device to raise the warp threads) and working in tandem with the bunched heddles to form the raised patterns on the fabric. The most prominent feature of this loom is its ability to weave texture of complex and large-scale patterns with frequent repeats.

17 漳缎织机
Zhangzhou Satin Loom

漳缎织机流行于清代，是我国古代花楼机中机械功能最为完善、机构最为合理、技术工艺最为成熟的花楼机。其主要技术特点体现在显花和应用起绒杆起绒的工艺上。这种独立式挂经装置的创新技术一直传承至今。

Prevailing in the Qing Dynasty, the Zhangzhou satin loom was a jacquard loom with the most perfect mechanical functions, the most reasonable structure and the most mature technological process. Its main technical characteristics are reflected in its techniques of weaving prominent patterns and raised fabric with a raising rod, as well as in its adoption of the innovative technology of the independent-style warp hanging apparatus, which is still in use today.

缂丝是一种中国独有的传统丝织工艺。它出现于唐代之前，在宋代达到鼎盛。缂丝业的中心在江苏省苏州市。织造时，织工用毛笔将画样的彩色图案描绘在铺好的经丝面上，以"通经断纬"为基本技法，以本色丝作经，各色彩丝作纬，根据纹样的轮廓或色彩的变化，采用不断换梭和局部回纬的方法织制图案。花纹正反两面如一，技艺精湛，格调高雅，是丝织工艺中最为高贵的品种。

The term Kesi, or K'o-ssu, denotes tapestry woven silk textiles. This traditional silk weaving technique, representing the most exquisite skill and quality of Chinese silk handicraft which is unique to China only. It first appeared before the Tang Dynasty and reached its height during the Song Dynasty, with Suzhou of Jiangsu Province being the center of the kesi industry. When weaving, workers first draw the colored pattern on the spread-out surface of the warp silk and then use undyed raw silk as warps and silk of varied colors as wefts to form a plain weave in the texture. According to the outline of the pattern or the change in colors, weaving is done on the k'o-ssu loom using the so called "thorough warps and broken wefts" method, which is characterized by the constant change of shuttles and local weft insertion, creating exquisite textiles with same patterns on both sides.

中国
古代科技之光
The Torch of Science
and Technology in Ancient China

19 轧花机 Ginning Machine

轧花机是一种去除棉籽的机械，出现于元代。轧花机主体为一个支架，架上部横向安装一个木轴、一个铁轴。铁轴在上，木轴在下，木轴右边装有曲柄，铁轴左边安装具有飞轮作用的木架。工作时右手转动曲柄，与曲柄相连的木轴随之转动，左脚踏动踏杆，使铁轴与木轴作反向运动，两轴相轧。左手将籽棉添入轴间，则棉花被带出轴前，棉籽落于轴后。

Used for deseeding cotton, the ginning machine first appeared in China during the Yuan Dynasty. The main part of the machine is a stand upon which are installed, in a horizontal way, a wooden roller (in the lower position) with a crank on its right side and an iron roller (in the upper position) with a wooden bar on its left side working like a flywheel. In the deseeding process, the worker turns the crank with his right hand to rotate the wooden roller while using his left foot to work the treadle to spin the iron roller. As the two rollers rotate in opposite directions, grinding against each other, unginned cotton is fed into the space between the rollers and deseeded.

20 细白麻布（文物） White Linen (Cultural Relic)

细白麻布是湖南省长沙马王堆一号墓出土的西汉织物，为平纹组织，是目前看到的最精细的汉代麻织品。

This white linen excavated from Mawangdui Tomb No. 1 in Changsha of Hunan Province. It was made by plain weave during the Western Han Dynasty and it is the most exquisite of its kind that has ever been found in China.

起绒锦是湖南省长沙马王堆一号墓出土的西汉锦，为提花丝织品，是同类织物中最早被发现的。它的花纹由绒圈组成，有浮雕状的立体效果。

Excavated from Mawangdui Tomb No. 1 in Changsha of Hunan Province, this Western Han Dynasty brocaded silk piece is the first of its kind that has ever been found in China. The raised decorative effect on this brocaded silk fragment was obtained by the incorporation of additional threads which have been drawn up into loops.

朱红菱纹罗是湖南省长沙马王堆一号墓出土的西汉的罗，为经提花织物，以粗细线条构成菱形图案，是汉代十分流行的高级丝织品。

Excavated from Mawangdui Tomb No. 1 in Changsha of Hunan Province, this gauze piece dates back to the Western Han Dynasty. Gauze was first produced in the Han Dynasty and is characterized by its fine open-work lozenge pattern formed by thick and thin lines. In the Han Dynasty, high-quality gauze such as this one was very fashionable.

烟色菱纹罗是湖南省长沙马王堆一号墓出土的西汉的罗，是汉代流行的高级丝织品。

Excavated from Mawangdui Tomb No. 1 in Changsha of Hunan Province, this gauze piece dates back to the Western Han Dynasty and was amongst the most fashionable high end silk products during the Han Dynasty.

乘云绣是湖南省长沙马王堆一号墓出土的西汉绣品。在中国古代，使用花色绣纹装饰单色丝绸的方法被广泛应用。这种西汉织物的绣地为绮，平纹地起斜纹花，图案为纵向的连续菱纹，再在菱纹内填横向菱花纹和对鸟。绣出的纹样单元较大，针法细腻流畅，为当时的高贵绣品。

Since the ancient times, embroidery in colored silks has been widely used in China to decorate a variety of monochrome silk textiles, sometimes with the addition of painted details. This Western Han Dynasty embroidery piece, featuring the theme of cloud-riding, was excavated from Mawangdui Tomb No. 1 in Changsha of Hunan Province. The base of the embroidery piece is damask, on which twill-weaving was applied to embroider continuous vertical lozenge patterns and within which horizontal designs of flowers and birds were filled. With its delicate and smooth needlework, this embroidery piece was a luxury at that time.

长寿绣是湖南省长沙马王堆一号墓出土的西汉绣品。绣品是在绢地上用四种丝线绣制而成的，针法为锁绣法。线条流畅，绣工细密，针脚整齐，为当时流行的高级绣品。

A Western Han Dynasty embroidery piece excavated from Mawangdui Tomb No. 1 in Changsha of Hunan Province, it was done on the silk base using silk threads of four different colors and with the lock-stitch method. Thanks to its smooth, dense and neat patterns, embroidery of this kind was highly fashionable at that time.

信期绣是湖南省长沙马王堆一号墓出土的西汉绣品。绣品在罗地上用多种丝线绣制而成，针法为锁绣法，图案纹样较小，做工精巧，为当时最讲究的一种绣品。

A Western Han Dynasty embroidery piece excavated from Mawangdui Tomb No. 1 in Changsha of Hunan Province, it bears sophisticated small patterns embroidered on the gauze base with serval types of silk threads and the lock-stitch method, making it the most exquisite embroidery at that time.

纺织刺绣
Textile and Embroidery

27 方格纹绣（文物）
Embroidery with Checkerboard Pattern (Cultural Relic)

方格纹绣是湖南省长沙马王堆一号墓出土的西汉绣品。绣品在黄褐色绢地上用墨绿色丝线绣成长宽各3厘米的斜方格，斜方格内为棕色和浅绿色的两种圈、点间行排列的纹样，十分精致。

A Western Han Dynasty embroidery piece excavated from Mawangdui Tomb No. 1 in Changsha of Hunan Province, its checkerboard pattern composed of 3 centimeters by 3 centimeters diagonal squares, embroidered on the yellow-brown silk base with dark green silk threads. Sophisticated small patterns, which are alternate motif of circle and dot in brown and light green colors, were embroidered within the diagonal squares.

28 西夏绣片（文物）
The Western Xia Embroidery (Cultural Relic)

宁夏回族自治区出土的西夏（1038—1227）绣片残片。西夏王朝也被称为西夏帝国，由定居于河西走廊和河套地区的党项族人建立。

The Western Xia Dynasty (1038–1227), also known as the Xi Xia Empire, or the Tangut Empire, was established by the ethnic group of Dangxiang round the area known as the Hexi Corridor, a stretch of the Silk Road.

Shown here is a fragment of the embroidery piece excavated in northwest China's Ningxia Hui Autonomous Region.

中国四大名绣

The Four Most Famous Embroidery Styles in China

中国的刺绣起源很早，相传"舜令禹刺五彩绣"。到秦汉时期（前221—220），刺绣工艺已能采用多种针法，绣出各种图案和动物形象。从宋代（960—1279）起，逐步形成刺绣准则，绣品除了用于装饰服装和日用品外，还出现了模仿书画的绣品。明清时期（1368—1911），刺绣技艺进一步提高，可将人物故事题材刺绣于服饰上。民间刺绣工艺也得到了发展，形成各具特色的地方体系，先后产生了湘绣、蜀绣、苏绣、粤绣四大名绣。

The origin of embroidery in China was very early. According to legend, Shun ordered Yu to make embroidery with five colors. During the Qin and Han Dynasties (221B.C.–A.D.220), the embroidery workers had already use multiple methods to embroider all kinds of patterns and animal images. Starting from the Song Dynasty (A.D.960–A.D.1279) and in addition to being used for decorating garments and articles of daily use, embroidery in China gradually developed into an art form specializing in the imitation of paintings and calligraphy. In the Ming and Qing Dynasties (A.D.1368–A.D.1911), with the development of the embroidery technic, themes such as figures and stories can be embroidered on garments. During that period folk embroidery skills had greatly improved, and showing more and more local characteristics. The most famous Chinese embroidery styles respectively originated in the City of Suzhou, as well as Hunan, Guangdong and Sichuan Provinces.

湘绣 《花木兰》
Hunan Embroidered Panel: Hua Mulan

湘绣为湖南地区的代表性刺绣，最先为民间刺绣，至清末艺术上臻于成熟，为中国四大名绣之一。湘绣的特点是用丝绒线绣花，色彩丰富鲜艳，绣件绒面花型具有真实感，形态生动逼真。左图所示绣品采用"双面全异绣"的绣法，以透明丝料为底，在正反两面绣制出两个不同主题、不同色彩和针法各异的人物形象。《花木兰》绣屏，以女英雄花木兰的传说为蓝本，一面绣的是"万里赴戎机，寒光照铁衣"的女扮男装、披坚执锐的英雄形象；另一面绣的是"当窗理云鬓，对镜贴花黄"的少女娇态。一面戎装，一面红装，形象生动，构思巧妙，绣工精湛。

Hunan embroidery began as a folk craft. It became artistically matured in the late part of the Qing Dynasty, ranking it one of the most famous embroidery styles in the country. Embroidered textiles from Hunan Province have been highly regarded in China. Hunan embroidery is characterized by its use of velvet silk threads to depict flowers, hence resulting in a sense of reality, varied colors and life likeliness. Traditional Chinese paintings are often used as its chief source. Featuring Hua Mulan, the legendary Chinese heroine, this example, called "double-faced discrepant embroidery", has been worked in satin stitch on a fine open-work gauze; it depicts the dual images of Mulan as a soldier and as a lady.

《望月》是一幅全异双面绣，一面为云鬓堆翠的仕女侧影，透过轻薄的竹帘，仰望明月；另一面是仕女微笑着的脸部侧影。这幅绣品构思巧妙，针法技艺高超，表达了仕女对美好爱情的向往。

This example is a double-faced discrepant embroidery. It depicts the front and back images of a beautiful lady gazing on the moon. With its delicate design and smooth needlework, this embroidery is ironically expressing the lady's yearning for love.

3

湘绣《锦上添花》

Hunan Embroidered Screen: Adding Flowers to
the Brocade

巨幅绣屏《锦上添花》是一件将绘画与巧绣融为一体的艺术珍品。它以白色软缎做底料，运用多种传统针法，用700多种彩色丝线，配以金线绣制而成。绣面上百花争艳，两只孔雀遥相呼应，形态逼真，给人一种呼之欲出的感觉。

This huge embroidered screen is a treasure of the Hunan style embroidery which combines the skills of painting and needlework. Different kinds of stitches are used to imitate hundreds of blooming flowers with gold thread and more than 700 kinds of colors. It features two peacocks so lifelike that they seem ready to step out of the screen.

蜀绣是以四川省成都为中心的代表性刺绣，起源于晋代（265—420），绣品色彩鲜艳，针脚整洁。本作品在织物基础上用彩线绣出了竹子和鲤鱼，象征着财富和繁盛。

Originated from the Jin Dynasty(A.D.265–A.D.420), Sichuan embroidery, made in and around Chengdu, the provincial capital of Sichuan, is known for its neat and tight stitches done with shiny and soft threads. Depicting bamboo trees and carps (Cyprinus carpio), this embroidered panel has been decorated in colored silks on a woven base fabric. In China, fish (Yu) is an ancient symbol of wealth and abundance.

古代科技之光
中国
The Torch of Science and Technology in Ancient China

5
苏绣《金鱼》
Suzhou Embroidered Panel: Goldfish

　　苏绣诞生于环境优美、物产丰富以及能工巧匠聚集的历史文化名城——苏州。苏绣沿袭1000多年前宋代的传统图案，利用精细有力的针法绣出风景、人物、花鸟等景物。画面秀丽素静、随物赋彩。在这幅作品中，绣工将一根丝线劈成四十八分之一的极细线，运用多种针法绣出金鱼精细的鱼鳍、鱼尾。绣工将上千个线结和线头巧妙地隐藏起来，绣出正反两面完全一样的金鱼图案。

Suzhou embroidery attests to the wealth of opportunity, environment, materials and craftsmen of Suzhou. Carrying on the tradition of the embroidered pictures of the Song Dynasty of 1,000 years ago, Suzhou embroiderers reproduce the brush-strokes of Chinese paintings and depict landscapes, personages, flowers and birds with strength or extreme delicacy. In this panel, the embroiderer employs silk threads only one forty-eighth as thick as that normally used, as well as a variety of stitches to depict the fine and delicate tails of the goldfish. Thousands of knots and thread ends are skillfully concealed so that the finished work can be viewed from both sides.

6
粤绣《西厢待月》
Guangdong Embroidered Panel: Waiting for the
Moonrise at the West Chamber

　　粤绣是广东地区的代表性刺绣。粤绣用线种类繁多，施针简单，色彩艳丽，往往红绿相间，用于渲染喜庆的气氛。这幅作品描绘了《西厢记》中的一个场景。在这幅绣画中，女主人公崔莺莺站在花园门口，静静地等待着她的爱人张生。绣品中人物神态生动，双眼晶莹闪烁，服饰飘逸，栩栩如生。

Originating from Guangdong Province, Guangdong embroidery is noted for its great variety of embroidery threads, wide range of subject matters, bright colors, rich designs and simple needlework. All of these elements combine to render the atmosphere of cheerfulness and excitement. This embroidered panel depicts a scene from the play Romance of the *West Chamber*. In this embroidered picture, Cui Yingying, the heroine, is shown standing by the garden gate, quietly waiting for her lover Zhang Sheng. Her eyes, measuring less than half a centimeter, were embroidered with dozens of threads of different colors which give them lifelike brilliance.

潮绣是粤绣的重要分支，至今已有 1000 多年的历史。潮绣以构图均匀、绣法特异、色彩浓艳著称。绣制时垫以棉花，用薄绢和纱覆盖，绣出花纹图案，再用金线绣制。这幅作品是典型的潮绣，作品中的牡丹花色彩艳丽。花瓶上绣有盘龙，旁边有一只狮子头顶装满寿桃的花篮。整幅绣品采用了高难度的垫绣技艺，用金银线和绒线绣制，画面宛如浮雕，富有立体感。

7
潮绣《玉堂春色》
Chaozhou Embroidered Panel: Jade Hall Teeming with Spring Hues

As an important branch of Guangdong embroidery, the Chaozhou style embroidery has a history of more than one thousand years. It is known for its uniform composition, unique stitches, rich and gaudy colors. In line with the design, the panel base is first padded with cotton and covered with thin silk and linen yarn, upon which the patterns are then embroidered. This embroidered panel is a typical Chaozhou embroidery piece; it has been worked in colored silks with silver and gilt metal threads on a relief woven foundation of cotton fabric; by using padding embroidery, a high-difficulty technique, enabled a raised decorative effect on this panel. The design on this panel depicts an elaborate arrangement of peonies accompanied by a lion supporting a basket of peaches. The embroidered detail on the bottle shows a four-clawed dragon chasing a flaming pearl among stylized clouds.

这件百子衣是明万历（1573—1620）孝靖皇后的刺绣短袄，长74厘米，对开襟，在罗地上绣出双龙、寿字。周身用金绒线绣入宝物、松树、竹子、梅花、石头和各种花草，并绣百子。百子生动活泼，各具姿态，有捕捉小鸟的，有捉迷藏的，有放风筝的，有头戴乌纱帽饰演官吏的，神态自如，惟妙惟肖，体现出刺绣艺术已达到很高的艺术境界。

Made during the reign of Emperor Wanli (A.D.1573–A.D.1620) of the Ming Dynasty, the original "hundred-children" coat belonged to Empress Xiaojing. It has been embroidered in gilt metal thread with a stylized foliate design which includes treasure, pine, bamboo plum and stone—symbolic references which stand for good fortune and longevity. The embroidered figures of children have been arranged informally. Some of them are shown catching birds, playing hide-and-seek and flying kites; some are shown wearing black gauze caps which is an accessory often worn by government officials. The coat attests that embroidery had already reached a high artistic level at that time.

清代龙袍
"Dragon Robe" of the Qing Dynasty

清代凤袍
"Phoenix Robe" of the Qing Dynasty

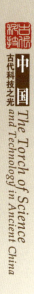

中国四大名锦

The Four Most Renowned Kinds of Brocade in China

　　锦上添花、繁花似锦，中文中通常以锦来描绘美好的事物和漂亮的景物。在众多丝织品中，锦缎是最为精美的品种。锦的历史悠久，花色品种很多，提花机的发明，使织锦工艺在唐宋时期发展成熟。元代是织锦技术的鼎盛时期，用金银线作纬线，织成富丽堂皇的织金锦，云锦便产生于此时。明清时期织物品种更加艳丽奢华，种类繁多，逐步形成了最为著名的四种锦，它们分别是：南京的云锦、四川的蜀锦、苏州的宋锦、广西的壮锦。

Brocade is usually used to describe beautiful things and wonderful scenery in Chinese culture. Of the great variety of silk fabrics, brocade is considered the most sophisticated. The invention of the jacquard loom in the Tang and Song Dynasties spurred the development of brocade, making even more gorgeous and complex patterns possible. The brocade reached its peak during the Yuan Dynasty, using gold and silver thread as weft to weave the magnificent Gold-thread brocade, also known as Yun brocade. The brocade of the Ming and Qing Dynasties were more gorgeous, luxurious and rich in varieties; and the four most renowned kinds of brocade were gradually formed. Those brocade are Song brocade of Suzhou, Shu brocade of Sichuan, Yun brocade of Nanjing and Zhuang brocade of Guangxi.

云锦

Nanjing Brocade

　　云锦为四大名锦之首，产于江苏省南京，是中国古代织锦最高水平的代表。云锦是在元代金锦基础上发展起来的，因纹样色彩灿烂、锦面绚丽似云而得名。

　　云锦可分为库缎、织金、织锦和妆花四大类。其中织金和妆花最为成熟，是云锦的代表。织金云锦使用纯金丝线来织造花纹，妆花云锦采用自由设计的方式创造出大胆而和谐的花色。云锦在最为繁盛的时期，主要供皇家使用。

The best among the four most renowned styles of brocade in China, Nanjing brocade is a traditional Chinese silk fabric which created in the Yuan Dynasty and flourished during the Ming and the Qing Dynasties. It is also called Yun (cloud) brocade because the varied designs and patterns of the fabric are comparable to beautiful clouds.

The major types of Nanjing brocade are palace, gold, tapestry and free-style. Gold and free-style are the most mature types and the main representative of Nanjing brocade. Gold brocade uses pure gold threads for the patterns while the free-style features free designs and bold yet harmonious colors. During its heyday, Nanjing brocade was mainly used by royalty.

蜀锦是四川地区生产的彩锦，已有 1000 多年的历史。蜀锦的品种花色甚多，属传统多彩提花丝织物。

With a history of more than 1,000 years and characterized by elegant and colorful patterns, Shu brocade of Sichuan, or Sichuan brocade, is produced in Sichuan Province. With a great variety of patterns and products, it belongs to the traditional type of multi-colored silk fabric woven in raised patterns.

宋锦起源于宋代，主产地为苏州。它采用 3 枚斜纹组织、2 种经纱、3 种色纬织成。宋锦的图案通常为格子藻井等几何纹框，纹框中有折枝小花，配色典雅和谐。主要用于书画装裱和官员服装。

Originating in the Song Dynasty and mainly produced in Suzhou, Song brocade is woven by using three twill textures, two kinds of warp and three kinds of weft. Geometric patterns, such as plaid caisson filled with small flowers, are usually featured in Song brocade whose color scheme is noted for its elegance and harmony. Representing the style of the Song Dynasty, Song brocade was mainly used for mounting painting and calligraphy scrolls, as well as for making uniforms worn by government officials.

　　壮锦出现于10世纪，是广西地区少数民族壮族民间的一种色彩绚丽的传统织物。它是以棉或纱作为经线，丝线作为纬线编织而成的精美织物。壮锦图案生动，结构严谨，色彩斑斓，充满热烈、开朗的民族格调。

Zhuang brocade incorporates a combination of silk and cotton fibers, with its weavers using cotton or linen yarn as warp and silk as weft. This traditional hand-woven craft first appeared in the 10th century and originates from Guangxi.

5

　　缂丝是中国丝绸艺术品中的精华。缂丝的织造方法不同于刺绣和织锦，这种工艺自宋代开始常与书画艺术相结合，通常以画为本或与画结合。织造时根据花纹的不同色彩，通经断纬，用若干小梭子挖织，织出十分细致的大型图案。

Kesi, or k'o-ssu represents the quintessence of Chinese silk art, its weaving method is different from that of embroidery and brocade. Starting from the Song Dynasty, this craft has often been associated with the art of painting and calligraphy, either using paintings as its chief source of designs or combining itself with paintings. It is characterized by the vertical slits which occur where one color ends and the next begins. Tapestry weave is well suited to the pictorial style of kesi and has been widely utilized to produce large-scale and exquisite designs.

　　《捣练图》是唐代画家张萱之作。它描绘了唐代城市妇女在制作丝绸过程中劳动操作时的情景。画中 12 个女性人物动作凝神自然，细节刻画生动。捣丝工有的手持杵棒，有的在挽袖子；理线工在精心地整理丝线；熨烫工倾斜着身体正在用力地拉平绸缎；烧炉工烧着炉子；一个好奇的人在四处张望。

Drawn by Zhang Xuan, an artist of the Tang Dynasty, Dao Lian Tu (Picture of Silk Processing) depicts a complete scene of silk processing and manufacturing. Plump and with exquisite manners, all of the 12 female figures in various postures and expressions are portrayed vividly and remarkably true to life: the silk pounders holding a stick or folding up sleeves; the silk menders adjusting the thread meticulously; the silk pullers bending backward forcibly; the furnace keeper warding off the heat and an spectator showing curiosity.

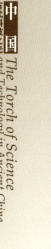
7 《捣练图》绢人

Silk figures Modeled to Show the Finishing Processes of Silk Making

与《捣练图》中描绘的场景相同，绢人们在忙着完成丝织物制作的最后工序。工人们将洗涤后的丝绸仔细熨烫，在熨斗里装的是燃烧的木炭。

As depicted in Dao Lian Tu (Picture of Silk Processing), the figures are busy with the finishing processes in the production of silk fabric. A length of silk is carefully washed and ironed. Note the inclusion of a charcoal burning stove and heating irons which were used to press the cloth.

　　公元前 139 年，汉武帝派遣张骞携带丝绸等物出使西域，开拓了横贯亚洲大陆直至地中海的古代贸易通道，之后海上丝绸之路也陆续开通。这举世闻名的陆上与海上丝绸之路对促进东西方文化交流和经济繁荣，对蚕、桑、丝、帛技术的西传起了巨大的作用。汉代的"丝绸之路"从关中的长安，穿过河西走廊和新疆的塔里木盆地，跨越帕米尔高原，然后经过如今的中亚各国，直达地中海东岸的港口，全长 7000 多千米。不仅商人沿着这条丝绸之路做丝、绢等贸易，朝廷也常常以中国的锦绣赠予外国的君王和使节。丝绸之路在历史上促进了亚、非、欧各国与中国的友好往来，增进了文化交流。

In 139 B.C., Emperor Wu of the Han Dynasty sent Zhang Qian, carrying silk and other goods, as a diplomat to visit the western regions. Zhang Qian's journey developed an ancient trade route across the Asian continent to the Mediterranean. The well-known overland and maritime Silk Road had greatly promoted cultural exchange and economic prosperity along the regions, and played a huge role in disseminating silkworm, mulberry and silk technology to the western world. From the Han Dynasty, the product was carried by the trade caravans over the 7,000 kilometer Silk Road which started from the capital Chang'an (today's Xi'an) of the Han Dynasty, ran through the Hexi Corridor, the Tarim Basin and the Pamirs, and reached the ports in the Mediterranean via the countries of

Central Asia. Via this overland route, the courts of the Han and Tang Dynasties sent silks as presents to their vassal princes, to officials in the border areas, and to foreign monarchs or ambassadors. Historically, both the overland Silk Road and its maritime counterpart have greatly facilitated the friendly interactions and cultural exchanges between China on one and the countries in Asia, Africa and Europe on the other side.

科古
技代

中国古代
科技之光

THE TORCH OF SCIENCE AND TECHNOLOGY IN ANCIENT CHINA

井盐开采

Well Salt Mining

　　中国是世界上最早生产井盐的国家。大约在 2300 年前，中国人就已开始通过人力挖掘的方法，开凿大口浅井，采集地下浅层盐卤了。1041—1048 年，井盐钻井技术出现了重大突破——宋代工匠发明了冲击式凿井法（又叫"顿钻凿井法"），并以这种方法凿成了一种称为"卓筒井"的小口径井。这种技术使得开采地层深处的盐卤资源成为现实。冲击式凿井法使钻井技术从人工挖掘发展为机械钻井，开创了人类机械钻井的先河。明清时期（1368—1911），井盐开采技术日臻完善，形成了钻井、修治井、采卤、输卤、制盐等一整套完善的生产工艺。1835 年，四川省自贡地区开凿出世界上第一口超千米的深井——燊海井，标志着古代钻井技术的成熟。

China is the first country in the world to have engaged in the production of well salt. About 2,300 years ago, the ancient Chinese started to use labor force to extract brine from large-mouthed shallow wells dug on the earth's surface. Between the years of 1041—1048, an important technological breakthrough in salt well drilling was made— the invention of the so-called impact method for sinking wells (also known as "the impulse-type well drilling method") by craftsmen of the Song Dynasty. With this technical method, they successfully drilled a small-mouthed well called "zhuo tong salt well", hence making it possible to extract the brine resources from the deeper layer of the earth's surface. Thanks to the invention of this improved method, well-drilling technology gradually evolved from manual digging into mechanical drilling, thus ushering in the age of mechanical well drilling for humanity. During the long period of 1368–1911 spanning the Ming and Qing Dynasties, well salt mining technology became increasingly consummate, as evidenced by the formation of a complete set of production process including well drilling, well maintenance, brine extraction, brine transfer and salt making etc. In 1835, people in the Zigong area of Sichuan Province succeeded in drilling the Shenhai Well, the first deep well in the world whose depth exceeded 1,000 meters. This milestone event represented the maturity of well drilling technology in ancient China.

　　碓架是井盐凿井设备。其结构是在木架上架设碓板，碓板中间固定，一端装钻头，另一端供工匠踩踏。凿井时，以人力踩动碓板，使钻头在井内作上下垂直运动，冲击并破碎井底岩石，钻出圆形的井眼。

Pestle Stand is a kind of salt well drilling equipment. The design of its structure was to set up pestles (levers) on the wood shelf. Fixed in the middle, the pestle had a drill bit on one side while the other side was treadled by the driller. In the process of drilling, the driller kept treadling the tip of the pestle so that the drill bit would make vertical up and down movements within the well to hit and smash the rocks in the bottom using its kinetic energy. By repeating this, the salt well could be dug deeper and deeper to form a round borehole.

　　天车为木制井架，是一种井盐生产的地面设备，最迟于 16 世纪出现，主要用于提取卤水，也用于钻井、淘井和修治井等。清代（1616—1911）天车的高度一般为 20～40 米，最高的达到 113 米。

Also known as "overhead traveling crane", such pithead frame appeared no later than the 16th century. It was a kind of ground equipment used for well salt production. Installed at saltworks, it was mainly used to extract brine; it could also be used to drill, to dredge or even to maintain a salt well. In the Qing Dynasty (A.D. 1616-A.D. 1911), the height of such pithead frames was usually between 20-40 meters, but the highest one could reach 113 meters.

3 凿井、捣泥模型 Model Showing Salt Well Drilling and the Lifting of Muck

凿井、捣泥是古代盐井锉井的主要工序。锉井时将锉井工具连接在碓架的碓板（踩板）前端，利用杠杆原理，以人力踩踏碓板后端，使锉井工具在井下反复运动，冲击岩层，使井不断加深。

捣泥就是将钻井时产生的碎石，利用地下水混合成泥浆后，用捣泥筒提出井口。捣泥筒由小于井径的竹筒做成，筒底悬一块熟牛皮构成单向阀。

Well drilling and the lifting of muck from the bottom of the well were the main work processes which involved in the process of drilling a salt well. When sinking a well, the digging tool was connected to the front end of the pestle set up at the pestle stand and, using the principle of lever, the back end of the pestle was manually treadled so that the digging tool would repeatedly make vertical up and down movements within the well to pound the rocks and deepen the well.

The muck resulting from the process of well drilling was mixed into a muddy sludge using underground water before being lifted out of the salt well. The lifting was done by means of a bamboo tube whose diameter was smaller than that of the well. A piece of dressed cowhide was attached to the bottom of the tube to constitute a unidirectional cock.

（1）（2）（3）（4）（5）

左图所示为五种盐井凿井工具。

（1）蒲扇锉：用于锉大口。

（2）马蹄锉：用于锉小口。

（3）鱼尾锉：锉井辅助工具。

（4）搰泥筒：用于提取泥浆。

（5）木柱：用于固井壁。

Showing here are five salt well drilling tools.

(1) Drill bit in the shape of a cattail-leaf fan: used to drill large boreholes.

(2) Fishtail drill bit: an auxiliary drilling tool.

(3) U-shaped drill bit: used to drill small boreholes.

(4) Muck lifting bamboo tube: an equipment used to lift the drilling-induced sludge out of the salt well.

(5) Wooden post: used to reinforce the wall of a salt well.

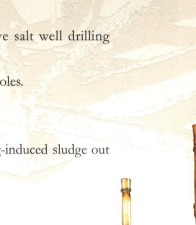

（1）　　　（2）　　　（5）

5 盐井修治工具
Salt Well Maintenance Tools

右图所示为五种盐井修治工具。

（1）铁五爪：用于打捞落石。

（2）倒插：修治井辅助工具。

（3）吊脚提须：用于打捞篾绳。

（4）怀胎扫镰：修治井辅助工具。

（5）偏肩：用于打捞井下落锉。

Showing here are five salt well maintenances tools.

(1) Iron Five Claws: used to retrieve fallen rocks.

(2) Inversed inserter: an auxiliary tool for well maintenance.

(3) Drop-point lifting bar: used to retrieve dropped-down bamboo rope (rope made of thin bamboo strips).

(4) Belly-bulged sweeping sickle: an auxiliary tool for well maintenance.

(5) Pianjian (literally, slanted shoulder): used to retrieve drill bits dropped into the salt well.

（3）

（4）

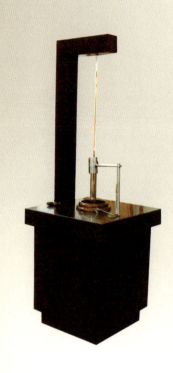

汲卤筒是一种用于提取井下卤水的工具。用小于井径的竹筒做成，筒底悬一熟牛皮构成单向阀。当汲卤筒向下放到井底时，卤水冲开阀门，进入筒内。当汲卤筒向上提升时，筒内卤水的压力将阀门关闭，从而将卤水提出井口。

左图为汲卤互动模型。

Featured in the model is a brine drawing pipe which, as its name suggests, was used to draw brine from salt wells. Pipes like this were usually made of bamboo tubes whose diameter was smaller than that of the wells. A piece of dressed cowhide was attached to the bottom of the pipe to form a unidirectional cock. When the pipe was lowered to the bottom of the salt well, the brine would force open the cock and enter into the pipe. When the pipe was hoisted, the pressure of the brine inside the pipe would close the cock, and the brine was thus lifted out of the salt well.

Shown here is an interactive model of brine drawing.

在钻井过程中，常常会遇到钻具掉入井内的事故。古代工匠在长期的井盐生产实践中，创造了一套打捞井下落物的技术。

右图为铁五爪打捞落石和偏肩打捞落铊的互动模型。

When sinking salt wells, drilling tools would occasionally fall into the wells by accident. In the long process of salt well production practice, the ancient craftsmen in China created a whole set of techniques for retrieving objects dropped down to the bottom of a salt well.

Shown here is an interactive model for experiencing the retrieval of fallen objects.

　　为了解决卤水和燃料异地分处的矛盾，中国人发明了一套完整的竹筹输卤工艺。用打通了竹节的大竹管，外缠竹篾，再敷上油灰，制成输送卤水的竹制管道。利用压差和连通器的原理，通过竹筹、筹窝、提水马车等设施，将卤水远距离输送到天然气产地制盐。

In order to solve the problem in well salt production which eaused by the long distance between the locations of brine and fuel, the ancient Chinese invented a complete set of technologies of transporting brine through a bamboo pipeline system, which became consummate during the Ming and Qing Dynasties. Big bamboo trunks, with their joints removed, were linked together to form a brine transportation pipeline, onto which thin bamboo strips were wound and then covered with a layer of putty. By making use of the principles of differential pressure and communication vessel and through a system of bamboo pipeline, bamboo "pipe nest" and water lifting cart etc. brine was transported over a long distance to the natural gas producing field for salt well making.

9 提卤站
Brine Lifting Station

提卤站是在输卤过程中用于提高水位的装置。

It was used to increase the water level during the brine transportation process.

10 枝条架浓卤塔
Brine Concentrating Wattle Tower

枝条架浓卤塔是一种井盐生产设施。它利用自然风力和日光蒸发卤水中的水分，达到浓缩和净化卤水、节约燃料、提高盐质的目的。

A facility for well salt production, it made use of natural wind and the thermal energy from the sun to evaporate the water from the brine in order to concentrate and purify the brine, save fuel and improve the quality of the well salt.

中国古代
科技之光
THE TORCH OF SCIENCE AND TECHNOLOGY IN ANCIENT CHINA

建

筑

Architecture

中国古代建筑是世界建筑宝库中的珍贵遗产。早在 7000 年前，中国人就发明了榫卯结构。中国古代建筑在长期的发展过程中，逐渐形成了一套以木结构框架式为主要特点，并结合夯土、斗拱、石雕、拱券及砖石工程等的传统建筑体系。在民宅、楼阁、陵墓、石窟、宫殿、寺院、古塔、园林、桥梁以及水利工程等建筑中都表现出独特的结构形式和艺术风格。

宋代时，中国的建筑风格更加典雅和精致。在砖、石、木结构技术方面有了新的发展。明清时期是古建筑繁荣期，今天我们仍旧能够看到当时的城市布局，品鉴那一时期的皇宫、花园和普通住宅的风格。

数千年来，中国古代建筑的传统技术一脉相承，自成体系，在世界建筑史上独树一帜。

The Chinese ancient architecture is a valuable legacy in the world treasury of architecture. As early as 7,000 years ago, the Chinese invented the mortise-and-tenon joint method, which was later to be used by ancient architects to erect imposing buildings out of numerous pieces of timber. In the long process of development, a traditional architectural system gradually took shape which is mainly characterized by wooden framework construction, coupled with the techniques of rammed earth, dougong, stone carving, arch and masonry engineering. The unique structural form and artistic style of this system are all embodied in the construction of civil residences, towers, tombs, stone caves, palaces, temples, ancient pagodas, gardens, bridges and irrigation works.

In the Song Dynasty, the Chinese building styles became more elegant and refined, and new progress was made in the techniques of bricks, stone and wooden frames. The last ancient architectural boom was witnessed during the Ming and Qing Dynasties when the styles of city layouts, imperial palaces, gardens and ordinary dwellings were finalized in the forms which we could see today.

For thousands of years, the traditional technology of ancient Chinese architecture has come down along a continuous line and formed a self-contained and unique style in the architectural history of the world.

河姆渡遗址距今 6900 年，是长江中下游已知年代最早的新石器时代遗址。遗址中发现大片干栏式建筑遗迹和大量带榫卯的木建筑构件。

这片木榫卯来自河姆渡遗址，是中国传统榫卯连接构造中的一个部件。榫卯是在两个木构件上采用凹凸结合的方式相互咬合，凸出的部分叫作榫，凹进去的部分叫作卯。榫卯结合，互相支撑，构成富有弹性的框架。尽管现在已经看不到那个时期的建筑物原本的模样，但是木榫卯构件证实了在河姆渡遗址中的干栏式建筑采用了榫卯结构。

The earliest vestige of Chinese architecture, this mortise or peg joint was found at Hemudu site (6900 years ago), the earliest known archaeological site of the Neolithic Age in the middle and lower reaches of the Yangtze River.

It is a component of the traditional mortise and tenon joinery, a flexible wooden frame system in which two pieces of wood are connected with rectangular pegs in rectangular holes. Despite the absence of any original structures of this date, the survival of this item has established the existence of the mortise and tenon structure in wooden stilt style buildings on this early site.

这是中国汉代流行的一种石刻建筑装饰，用于祠堂或墓室等建筑装饰。作品构图、造型和线条质朴生动，是中国珍贵的艺术遗产之一。

Stone carving was an essential feature of ancestral temples and funerary architecture during the Han Dynasty and often incorporated decorative details relating to daily life, social customs and mythological tales. This representation is typical of Han tomb sculpture and is distinguished by its simple and lively composition, shape and lines, making it an important part of the invaluable artistic heritage of China.

建筑
Architecture

3 《车马过阙》画像砖
Stone Relief

《车马过阙》画像砖制于东汉，出土于河南省许昌市。砖上刻画了马车穿过城门的画面。

Made in the Eastern Han Dynasty, it was unearthed in the city of Xucheng in Henan Province. This stone relief depicts a horse-drawn cart passing through a gate tower.

4 《上人马食大仓》画像砖
Stone Relief

《上人马食大仓》画像砖制于东汉，出土于河南省许昌市。砖上刻画了马车穿过城门的画面，并刻有"上人马食大仓"。

Made in the Eastern Han Dynasty, it was unearthed in the city of Xuchang in Henan Province. This stone relief depicts a horse-drawn cart passing through a gate tower. It also bears the inscriptions "Shang Ren Ma Shi Da Cang", or to live a luxury and privileged life.

5 长城砖
The Great Wall Brick

长城是中国古代工匠在建筑和工程领域的巅峰巨制，是世界八大奇迹之一。此砖为长城砖，根据砖印记载，其为明代烧制。

This brick was removed from the ruins of the Great Wall, which is among the most magnificent construction engineering projects undertaken by the ancient architects of China and one of the eight wonders of the world. The mark on the brick indicates it was made in the Ming Dynasty.

瓦出现于西周初期，嵌固在屋顶表面，用于屋顶防水。瓦分筒瓦和板瓦，考究的房屋用筒瓦盖瓦垄，一般房屋全部用板瓦。

Earthen roof tiles first appeared during the early Western Zhou Dynasty; they were placed and fixed on top of a building in order to make it waterproof. Roof tiles were often divided into arched tiles and pan tiles. Generally speaking, in ancient times luxurious buildings used arched tiles as roofing whereas ordinary domestic houses used pan tiles instead.

瓦当出现于西周，装饰于筒瓦头部。檐口处的筒瓦头称为瓦当，板瓦头称为滴水。秦汉时，瓦当的装饰风格明显，多以动物形象为主，如鹿纹、鸿雁、鱼等。图案反映了人们祈福求祥的心理，也有驱邪的画像或使用墓主人的形象。典型的是四种神话动物：青龙、白虎、朱雀、玄武。青龙能够呼风唤雨，为四神之首。

First appeared in the Western Zhou Dynasty, eaves tiles on traditional Chinese buildings were placed at the eave end of a row of roof tiles to form a tidy edge. Some of them were decorated with characters designating the building or tomb for which they were intended.

Traditionally, eaves tiles used for arched roof tiles are called "wa dang" while those used for pan roof tiles are referred to as "di shui", or drip-tiles.

Eaves tiles dating from the Qin and Han Dynasties exhibited a number of different decorative devices. Most of them were decorated with animal imagery such as deer, wild goose and fish etc., reflecting people's wish for good luck and peace. Images believed to be able to drive away evil spirits were also used, the most typical of them being four mythical animals, namely, blue dragon, white tiger, rosefinch and black tortoise. Since according to legends, blue dragon could summon wind and rain, it was the top-ranked one among the four mythical animals.

脊兽是中国古代重要建筑屋脊上的特殊装饰件。在屋顶的正脊、垂脊、岔脊之上，置有大小不等、形状各异的脊兽。正脊两端叫"正吻"，在垂脊和岔脊的末端置有一队小兽，领头的是一个仙人，而后依次为龙、凤、狮、天马、海马、押鱼、獬豸、斗牛和行什。脊兽的安装数量依建筑等级高低和规模大小而定。

Glazed ceramic ridge tiles modeled in the form of mythical figures and auspicious animals added an important structural and decorative element to the roofs of significant buildings, such as the imperial palace, temples, altars and palaces of princes etc. Made in different sizes and shapes, they usually were placed on the principal ridge, the vertical ridge and the branch ridge according to their traditional significance.

The "Zhengwen" or "great animals" were arranged at both ends of the principal ridge, facing inwards and biting into the ridge-beam. Smaller figures were placed in line on the vertical and branch ridges; and led by the figure of a fairy, they were arranged in the order of dragon, phoenix, lion, the heavenly steed, seahorse, fish, "xiezhi", the fighting bull and "hangshi".

The number of such ridge tiles used on each building was determined by its size and status.

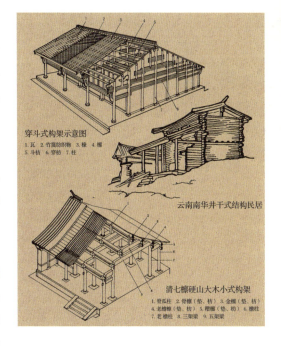

琉璃瓦在北魏（386—534）时期开始使用，唐代普及，宋代盛行。琉璃是一种光亮、不渗水的黄、绿、蓝等颜色的彩釉。琉璃砖瓦是将琉璃涂于陶砖瓦表面烧制而成的。

Pottery tiles and bricks colored on the surface with yellow, green or blue glaze are known as glazed tiles and glazed bricks. They are shining and waterproof. Glazed tiles and bricks first started to be used in the Northern Wei Dynasty (A.D. 386–A.D. 534), but their use required special permission from the emperor and was largely limited to royal and public buildings. For ordinary mortals, they had to make do with plain tiles and bricks.

10

中国古代木构架结构

Illustration of Wooden Framework Construction in Ancient China

中国古代建筑以木结构为主要的结构方式，有抬梁式、穿斗式和井干式三种。抬梁式使用范围较广，在三者中居于首位，春秋时已出现，唐代发展成熟。穿斗式在中国南方诸省普遍使用，这种木结构技术在汉代已相当成熟。井干式结构多在少数民族地区使用。

穿斗式构架示意图
1. 瓦 2. 竹篾纺织物 3. 椽 4. 檩
5. 斗枋 6. 穿枋 7. 柱

云南南华井干结构民居

清七檩硬山大木小式构架
1. 脊瓜柱 2. 脊檩（垫、枋）3. 金檩（垫、枋）
4. 老檐檩（垫、枋）5. 檐檩（垫、枋）6. 檐柱
7. 老檐柱 8. 三架梁 9. 五架梁

The ancient Chinese architects preferred wooden framework construction, which was divided further into three types, namely, the post-and-lintel construction, the column-and-tie construction and the log cabin construction. Of the three types of construction, the post-and-lintel construction, dating far back to the Spring and Autumn Period and becoming maturely developed in the Tang Dynasty, was the foremost and most commonly used technique. The column-and-tie construction, already well-developed during the Han Dynasty, was widely applied in the provinces south of the Yangtze River. The log cabin construction was mainly adopted in areas inhabited by the ethnic minorities.

斗拱是中国传统木构架体系建筑中独有的构件，斗是斗状方形木块，拱是弓形的短木，拱架在斗上，纵横交错叠加，逐层向外挑出，形成上大下小的托架。周朝时，已经有在柱上安装坐斗、承载横枋的方法。汉代时，成组的斗拱已大量用于宫殿、寺庙等高级建筑中；唐宋时期，斗拱较粗壮，样式统一；明清时期，斗拱的体量逐步缩小，更加突出装饰性。

斗拱主要用于柱顶、额枋、屋檐或构架间，在木构架结构中起着承上启下、消能减震的作用，它既传递荷载，增加出檐深度，又是精美的装饰。

One of the most unique features of traditional Chinese architecture is this complex jumble of wooden corbel brackets known as "dougong". Each set of "dougong" consists of tiers of beams called "gong", cushioned with trapezoidal blocks called "dou". The "dou" and "gong" crisscross and extend outward layer by layer forming an inverted pyramid structure. In the Zhou Dynasty, there was already a way to install the "dou" on the column to load the crossbeam. In the Han Dynasty, a group of "dougong" had been used in a large number of high-grade buildings, such as palaces and temples. During the Tang and Song Dynasties, the "dougong" were more robust and unified. In the Ming and Qing Dynasties, the volume of the "dougong" gradually decreased, but its ornamental function was more prominent.

"Dougong" was mainly set on top of the columns to support the beams within and eaves without, acting as a link between the upper and lower parts of the wooden frame structure. Its sophisticated design is not only an exquisite decoration, but also can transfer the load, dampen earthquake energy and increase the depth of eaves.

中国古代科技之光 *The Torch of Science and Technology in Ancient China*

　　唐代都城长安城建于 7 世纪，是当时世界上规划最合理、宫殿最完美、规模最宏大的都市。长安城布局合理，全城由郭城、宫城、皇城构成，面积达 84 平方千米。宫城位于城中心北部，皇城在宫城南面，是中央官署区。郭城里有南北并列大街 14 条，东西平行大街 11 条，这些街道将全城分成 108 个里坊，东西两市各占 2 坊，全城的商店、作坊都设于此。

Built in the 7th century, Chang'an (today's Xi'an) was the national capital of the Tang Dynasty. It was the most rational planning, the largest scale city with the most perfect palace. Planned in perfect order and with a collection of refined palaces, the entire Chang'an City consisted of the outer city (residential and industrial and commercial areas), the palace city (district of imperial palaces) and the imperial city (district of government offices), covering an area of 84 square kilometers. The imperial city, temples and government offices were all built facing south on the city's north-south axis. The 108 squares (lifang) formed by a network of 11 north-south streets and 14 east-west streets made the layout of the city look like a gigantic chessboard. Each occupying two squares, the East Marketplace and the West Marketplace were home to all the stores, workshops and factories in the city.

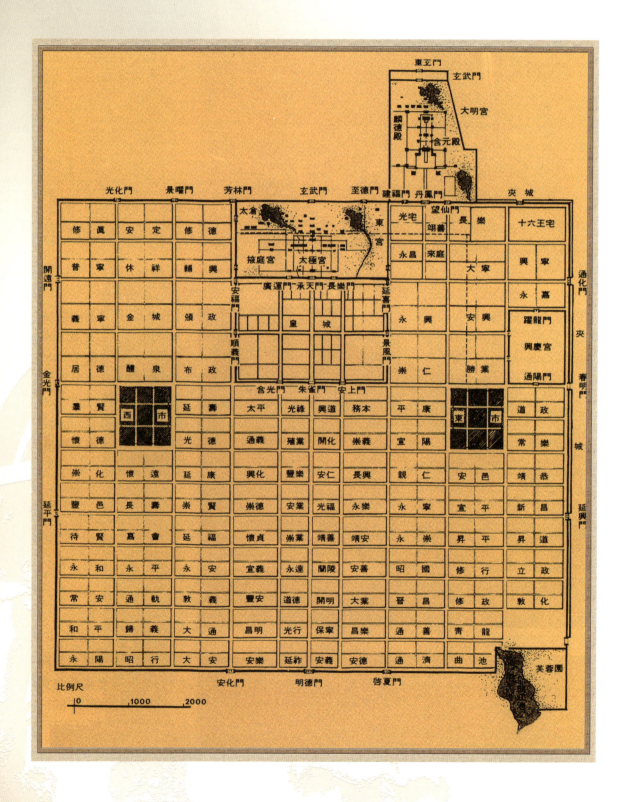

　　佛光寺位于山西省五台县豆村东北，建于北魏孝文帝时期（471—499），并于唐大中十一年（857）重建。佛光寺大殿是现存唐代殿堂型构架建筑中最古老、最典型、规模最宏大的一例。粗壮的柱列、简朴的门窗、深邃的层檐、舒展的屋顶，形成浑朴雄壮的外观。大殿共采用七种形制的斗拱，是现存建筑中挑出层数最多、挑出距离最远的建筑。大殿荟萃了唐代建筑、雕塑、书法和绘画四种艺术于一堂，具有极高的历史和艺术价值。

Located in the northeast of Doucun Village in Wutai County of Shanxi Province, the Foguang Temple was built during the era of Emperor Xiaowen of the Northern Wei Dynasty (A.D.471–A.D.499) and reconstructed in the 11th year (A.D.857) of the reign of Dazhong in the Tang Dynasty. With big and strong columns, simple doors and windows, deep multiple tiered-eaves, extended roofs and forceful exterior, the main hall of the temple is the oldest and largest existent palatial architecture of wooden framework construction most typical of the Tang Dynasty. As many as seven types of dougong techniques were applied in the construction of the hall, making it a building with the most and the longest stretching out tiers of eaves. Within the hall is a concentration of Tang Dynasty art in the forms of architecture, sculpture, calligraphy and painting. As such, the hall has extremely high historical and artistic value.

建筑
Architecture

14 永乐宫
The Yongle Palace

　　永乐宫是元代道观建筑典型，始建于 1247 年。现位于山西省芮城县龙泉村。现存的永乐宫为一门三殿，一门为龙虎殿，也叫无极门，三殿为三清殿、纯阳殿、重阳殿，依中轴线向北排列。永乐宫建筑规模宏大，布局疏朗，斗拱重叠交错，藻井结构精巧，主要殿堂的内部绘有精美的壁画，是现存较完整的元代建筑之一。

Located at the Longquan Village, the north of Ruicheng County in Shanxi Province, the Yongle Palace was built in 1247. As a typical Taoist temple, Yongle Palace has one entrance hall and three major halls. The entrance hall is known as the Longhu Hall of Wuji Hall. The three major halls are the Sanqing Hall, the Chunyang Hall and the Chongyang Hall, in which there are beautiful and exquisite murals. The entire palace, which is relatively complete and well preserved, has a reasonable layout, a complex overlapping Dougong system, a fine caisson structure and was intricately designed, making it a magnificent and majestic architecture of the Yuan Dynasty.

独乐寺始建于唐代，于辽圣宗统和二年（984）重建，坐落于天津市蓟州区。寺内观音阁是一座三层木构楼阁，通高23米，是中国现存最早的木结构楼阁式建筑。独乐寺历经千年，经历28次地震，仍安然无恙。

Located inside the Jizhou district of Tianjin, the Dule Temple was reconstructed in the 2nd year (A.D.984) of the reign of Tonghe in the Liao Dynasty on the foundations of an established temple complex first built in the Tang Dynasty. A three-stored wooden structure with a height of 23 meters, the Guanyin Pavilion inside the temple is one of the best preserved wooden buildings in China. Over the past millennium since its construction, it survived twenty-eight earthquakes.

建筑
Architecture

16 大雁塔
The Great Wild Goose Pagoda

大雁塔建于唐永徽三年（652），坐落于陕西省西安市慈恩寺内。原高5层，后扩建为9层，历经多次修缮和改造后，现在的大雁塔共7层，塔高64.5米。大雁塔呈方形角锥状，用青砖仿木结构建成，各层四面设有砖砌拱门，造型简洁，气势雄伟。它是中国现存最早的楼阁式塔，也是中国古塔建筑的优秀典型。

Located inside the Ci'en Temple complex in the City of Xi'an of Shaanxi Province, the Great Wild Goose Pagoda was built in the 3rd year (A.D.652) of the reign of Yonghui in the Tang Dynasty. It originally had five storeys but was later extended to nine. After many phases of construction and deconstruction, the pagoda now has 7 storeys and is 64.5 meters high. Each storey of the pagoda has arched gates on four sides. The pagoda was built with brick wood structure, takes a square pyramid form with simple, elegant and spectacular modeling, making it the oldest and largest existent Tang Dynasty tower style brick pagoda, as well as a typical example of ancient traditional Chinese architecture.

　　小雁塔建于唐景龙元年至三年（707—709），坐落于陕西省西安市荐福寺内。原塔有 15 层，明代时，遭遇多次地震，塔身震裂，塔顶 2 层被毁，现存 13 层。塔高约 43.3 米，后经历二次修缮，基本保存了原有的格局，小雁塔塔形秀丽，是中国早期方形密檐式砖塔的典型建筑。

Located inside the Jianfu Temple in the City of Xi'an of Shaanxi Province, this pagoda was constructed between the 1st to 3rd years (A.D.707–A.D.709) of the reign of Jinglong in the Tang Dynasty. It was badly damaged in the earthquakes occured in the Ming Dynasty, and the two topmost of its 15 storeys collapsed. After two repairs, the original structure was basically restored. Despite the loss, this brick building, with its remaining 13 storeys, still stands a towering height of 43.3 meters. It is representative of Chinese Pagoda architecture.

佛宫寺释迦塔
The Sakyamuni Pagoda of the Fogong Temple

佛宫寺释迦塔又称应县木塔，建于辽清宁二年（1056），位于山西省应县佛宫寺内。塔呈平面八角形，外观为 5 层 6 檐，内设有暗层，共 9 层。塔高 67.31 米，底层直径 30.27 米。整个建筑采用 60 多种斗拱技术。900 多年来，木塔经历了 10 余次地震毫无损坏。是中国现存最古老、最大、最高的多层木结构建筑，也是世界现存最高的古代木结构建筑。

Located inside the Fogong Temple complex in Yingxian County of Shanxi Province, this all-wood pagoda is sometimes referred to as "the wooden pagoda of Yingxian". The octagonal pagoda was built in the 2nd year (A.D.1056) of the reign of Qingning in the Liao Dynasty and has five storeys (nine storeys if the mezzanines are also included) and six levels of eaves, standing 67.31 meters high with a diameter of 30.27 meters at its ground level. The entire structure employed as many as over sixty varieties of dougong (corbel bracket system) and has withstood more than ten earthquakes over a period of some 900 years after its construction. As the oldest, largest and tallest multi-storeyed wooden architecture still in existence in China, as well as the tallest of its kind in the world, the achievements embodied in the pagoda in terms of architectural structure, technology and art, make it an important object in studying the ancient Chinese history of architecture.

　　紫禁城，又称故宫，始建于明永乐四年（1406），历经 14 年建成。是明清两代的皇宫。它坐落在北京城中心，南北向 961 米，东西向 753 米，占地面积 72000 平方米，是我国现存最大最完整的古建筑群。

　　故宫的全部建筑分为"外朝"和"内廷"两大部分，有大小宫殿 70 余座，房屋 9000 余间。外朝以太和殿、中和殿、保和殿三大殿为主，太和殿是当时规格最高的建筑，也是古代最大的木结构建筑。三大殿两侧有文华、武英两组宫殿，是皇帝举办重要国家仪式、会见朝廷官员的场所。内廷以乾清宫、交泰殿、坤宁宫为主，两侧有东西六宫和宁寿宫、慈宁宫等宫殿，是皇帝和皇后、嫔妃以及宦官的生活住所。宫殿群外有 10 米高的城墙和 52 米宽的护城河，城墙四面有门，四角设角楼。紫禁城主要建筑按照其功能和地位分布于中轴线上，保持对称布局。建筑普遍采用红色墙、柱子作为装饰，屋顶上覆黄色琉璃瓦。紫禁城代表了中国古代建筑群布局设计、艺术风格和建筑工程成就的最高水平。

Located at the heart of the central axis of Beijing, the Forbidden City served as the imperial palace for both the Ming and the Qing Dynasties. It took fourteen years to complete the entire complex. Construction of the Forbidden City, on a land area of 720,000 square meters, was started in the 4th year (A.D.1406) of the reign of Yongle in the Ming Dynasty.

More than 70 palaces of different scales and over 9,000 rooms inside the Forbidden City were divided into an "outer court" (public front section) and an "inner courtyard" (private rear section). The outer court mainly consists of the Hall of Supreme Harmony, the Hall of Perfect Harmony, and the Hall of Preserving Harmony and with the Hall of Literary Glory and the Hall of Martial Valor along both sides. The Hall of Supreme Harmony, located in the center of the Imperial Palace, was the highest-ranking building at that time and the largest wooden building in ancient times. The inner courtyard has the Palace of Heavenly Purity, the Hall of Union and the Palace of Earthly Tranquility as main buildings, and with the Six Eastern Palaces, Six Western Palaces, the Palace of Tranquil Longevity and the Hall of Consolation of Mothers on the two sides. All the main buildings were situated along the central north-south axis. The emperors undertook important state ceremonies and met their court officials in the outer court whereas the inner courtyard served as domestic quarters for the emperors, empresses, concubines and eunuchs. Encircled by a 10-meter high city wall and surrounded by a 52-meter wide moat, the Forbidden City measures 961 meters from north to south and 753 meters

建筑

Architecture

from east to west, making it the largest and best preserved ancient architectural complex in China. There are entrances on four sides of the city wall and at each corner of the wall, there is a watchtower. Red walls, decorated pillars and yellow glazed tiles are just commonplaces in the Forbidden City. Adopting symmetrical arrangements of courtyards which are divided according to their functions and significance, the Forbidden City represents the highest level of layout design, artistic style and construction engineering achievements of ancient Chinese architectural complex.

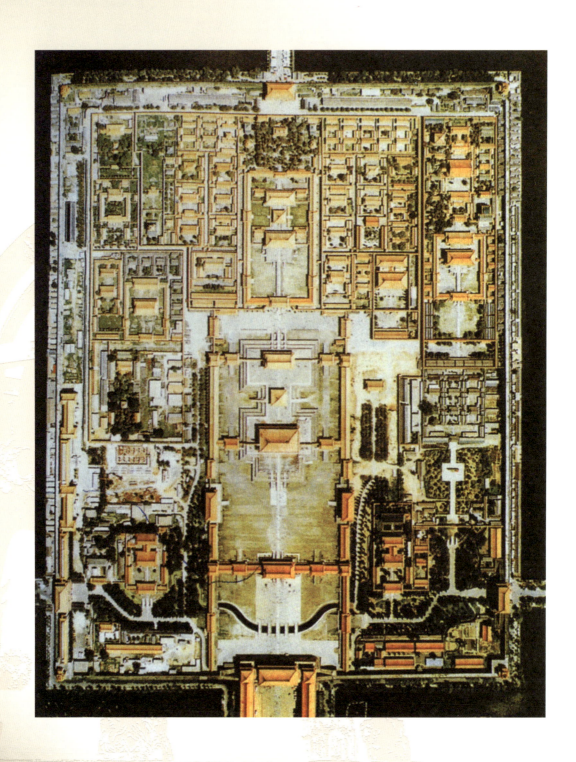

赵州桥建于隋开皇十五年至大业元年（595—605），位于河北省赵县南门外的洨河上，是古代中国南北交通的主要干道上的石桥。赵州桥由匠师李春主持建造，是世界上最早的石拱券敞肩桥。桥长50.82米，宽9.6米，跨度为37.37米，呈大弧形横跨洨河。赵州桥经历过洪水、地震等自然灾害的袭击，历经1300多年，至今完好无损，反映了中国古代高超的桥梁建筑技术。

China's extraordinary rich legacy in bridge construction is best represented by the Zhaozhou Bridge over the Jiaohe River in Zhaoxian County of Hebei Province. Built in the Sui Dynasty between A.D.595 and A.D.605, the stone bridge served as a key link connecting the north-south thoroughfares of ancient China. It mainly consists of a single segmental arch. Four smaller arches, two at each shoulder of the main arch, help to support the roadway; and in the case of a flood, act as spillways. The stone bridge, the construction of which was supervised by a craftsman known as Li Chun, is the first of its kind in the world to have adopted the open spandrelled arch design. It has a length of 50.82 meters and a width of 9.6 meters, segmentally spanning 37.37 meters over the Jiaohe River. Having withstood the test of floods, earthquakes and other natural disasters over the past 1,300-plus years, the bridge still remains intact, which is a testament to the superb achievements in bridge construction engineering in ancient China.

建筑
Architecture

21 《营造法式》
Yingzao Fashi (The Methods of Architecture)

《营造法式》于宋崇宁二年（1103）颁行，由李诫编修。全书共分5部分，357篇。该书是制定各种建筑设计、结构、用料和施工的建筑规范的著作，对宋代以后的建筑技术发展起到了促进作用。

Written by Li Jie, a noted architect living in the Northern Song Dynasty, *Ying zao Fashi* was published in A.D.1103 for the management of the construction of the imperial palace, temples, government buildings and official mansions etc. It is composed of five parts in 357 chapters and is the world's first building code and set of regulations. The purpose of the book was to standardize the designs and structures, as well as the use of building materials and methods of construction for all kinds of architectural projects. As a de-facto encyclopedia for traditional Chinese architecture, the book played an important role not only in standardizing the construction of architectures in ancient China but also in promoting the development of architectural technologies after the Song Dynasty.

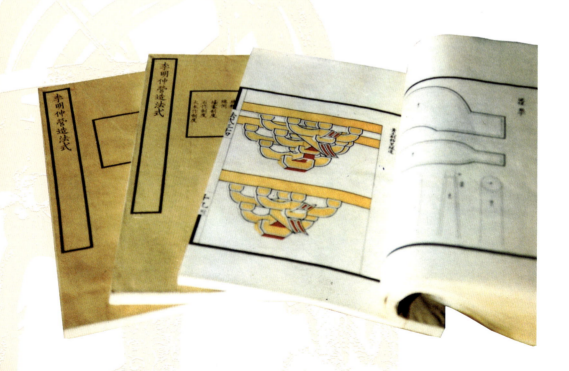

　　《清明上河图》由北宋张择端绘制，反映了当时京都汴梁（今河南省开封市）近郊汴河两岸清明时节繁荣的社会生活、建筑和地理风光。《清明上河图》采用中国传统绘画独特的写实手法将物象提炼升华。该卷长 528 厘米，高 24.8 厘米，为我国古代风俗画类型的杰作。

A genre painting by Zhang Zeduan of the Song Dynasty, Riverside Scene at the Qingming Festival depicts in detail the generally thriving urban scenes and prosperous atmosphere, the heavy traffic, the various activities taking place and their vitality, architectural styles and geographical landscape etc. in the City of Bianliang (present-day Kaifeng), capital of the Northern Song Dynasty, during the day of the Qingming Festival, or Tomb-Sweeping Day, when people visit ancestral tombs. In the form of a roll, the painting is 528 centimeters in length and 24.8 centimeters in height. As a priceless treasure in the history of Chinese painting, this painting speaks eloquently of the highest level of skill attained in popular art.

23

平江图

Rubbing Piece of the Layout of Pingjiang

下图是南宋绍定二年（1229）刻制的平江图（今苏州市）石碑拓片，碑高 2.76 米，宽 1.41 米，是最早的街巷制城市平面图。左图中北部居住区都是东西向长巷，巷内建宅，巷外街道为商业街。各巷口有二柱一楼的牌坊、沿街设店、跨街建坊是街巷制城市特点；南街道多为网状方格。此图是中国传世最早、保存最好的城市平面图。

In the Song Dynasty, Suzhou was called the Prefecture of Pingjiang. Copied from a stone stele carved in the 2nd year of the reign of Shaoding (A.D.1229) in the Southern Song Dynasty and measuring 2.76 meters high and 1.41 meters wide, this rubbing

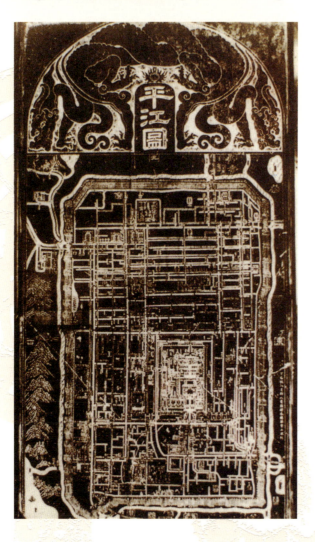

piece illustrates the symmetrical arrangement of buildings within the city and also shows the Grand Canal which supplied it with fresh water. From the rubbing, we can see that the northern part of the city consisted of east-west running long lanes within which residences were built, while the streets outside the lanes served as commercial streets which were lined by shops on both sides. In contrast, the southern part of the city was crisscrossed by a chessboard-like street system in the absence of combined horizontal long lanes. As the earliest and the best existent city layout plan, the rubbing piece provides historians with important sources for the study of city planning in ancient China.

万里长城是中国古代在不同时期为抵御游牧部落和别国的侵袭而修筑的规模浩大的军事工程。长城始建于公元前5世纪，由秦、燕、赵三国分别建造。公元前221年秦始皇统一中国后，将秦、赵、燕三国的长城连为一体，形成最早的长城。从秦以后，历代对长城的修筑持续了1000多年，长城的规模不断扩大，到明代末年时达到顶峰，也就是我们如今看到的规模。

明长城以砖石筑造，形如巨龙，总长达8851.8千米。它东起辽宁省虎山，西至甘肃省嘉峪关，途中跨越山脉和沙漠，经过辽宁、河北、天津、北京、山西、内蒙古、陕西、宁夏、青海、甘肃等省、自治区和直辖市的156个县域。长城是世界建筑史上的奇迹之一。

The Great Wall has its origins in the walls of pounded clay built up during the 5th century B.C. by the states of Qin, Zhao and Yan along the northern and western defenses to prevent the invasions of the nomadic tribes. After King Yingzheng of the State of Qin unified China in 221 B.C. and established the Qin Dynasty, the existing sections were joined together and extended to form the Great Wall. Over the following centuries, the Great Wall was further extended and reinforced on a continuous basis, attaining its final form as we see today during the Ming Dynasty.

Like a majestic dragon, the Great Wall winds 8851.8 kilometers over mountain ranges and deserts from Hushan Town, Liaoning Province westward to Jiayuguan Pass in Gansu Province, passing through 156 counties such as Liaoning, Hebei, Tianjin, Beijing, Shanxi, Inner Mongolia, Shaanxi, Ningxia, Qinghai, Gansu and other provinces, municipalities and autonomous regions. Arguably, it is one of the wonders in the world history of architecture.

中国古代
科技之光

THE TORCH OF SCIENCE AND TECHNOLOGY IN ANCIENT CHINA

陶

瓷

Ceramics

　　陶瓷的生产是文化发展的标志之一。中国的陶瓷生产可以追溯至7000多年前的新石器时代。当时，居住在黄河沿岸、长江中下游以及东部沿海地区的散落居民普遍用手工制造和使用各种陶器。逐渐发展起来的制陶工艺，特别是在龙山文化后期发明的陶轮，为瓷器的诞生奠定了坚实的物质和技术基础。

　　约3000年前的商代，人们发现烧制之前在陶坯表面挂一层硅酸釉会使成品的表面特别光滑和富有光泽。于是，最早的瓷器"原始青瓷"就诞生了。3—5世纪，中国出现了真正意义上的瓷器。随后的几个世纪，制瓷技术趋于成熟，并达到相当高的造诣。10—19世纪，瓷器制造在成型工艺、釉料、烧制方法及绘画和上釉种类等各方面都达到了顶峰。

　　中国瓷器的发展经历了从青瓷到白瓷再到彩瓷的几个阶段，逐渐达到了极高的艺术水准，在中国技术、工艺和文化发展史上谱写下瑰丽的篇章。在英文中"瓷器"（china）一词已成为"中国"（China）的代名词。16世纪以来的全球贸易，使瓷器成为人类贸易史上最早的全球化商品之一。

The production of ceramics is one of the marks of cultural development. In China, it can be traced back more than 7,000 years in the Neolithic Age, when hand-shaped pottery utensils were commonly made and used by people living in the scattered human settlements on the banks of the Yellow River, in the lower and middle reaches of the Yangtze River, and along the eastern sea coast. The gradually developed pottery-making techniques, especially the invention of the potter's wheel in the later period of the Longshan Culture, laid a solid material and technological foundation for the birth of porcelain.

About 3,000 years ago during the Shang Dynasty, it was discovered that coating the surface of a greenware piece with silicate glaze before kilning it would give the baked wares an especially smooth and brilliant finish. Consequently, the earliest form of porcelain known as "proto-celadon" came into being. In the period from the 3rd to 5th centuries, the first porcelain in its true sense appeared in China; and in the ensuing centuries therefrom, porcelain-making gradually matured and reached a high degree of perfection. From the 10th–19th centuries, porcelain was at its peak in molding techniques, glazing materials, firing methods and variety of painting and glazing.

From the greenish glazed celadon to white porcelain and then to colored glazed porcelain, the development of porcelain in China has experienced several stages. Artistically the Chinese porcelain has reached a very high level, making it a splendid chapter in the Chinese annals of technology, craft and culture, as well as a synonym of China. And as a result of the global trade after the 16th century, it also became one of the first globalized commodities in human trade history. the developments of ceramics in China .The development of porcelain had evolved in three stages that are celadon, white porcelain and painted porcelain. Chinese porcelain had reached the very high artistic level, and been continually introduced to the world. It has been widely praised. The invention of Chinese ceramics, which displayed the great creativity of China, has made outstanding and superior contributions to the world civilization.

陶器 *Pottery*

中国是世界上早期发明陶器的国家之一，至今已有 7000 ~ 8000 年的历史。新石器时代陶器种类很多，有灰陶、红陶、彩陶、黑陶、白陶、釉陶等。陶器之所以呈现出不同的颜色，主要是因为制陶原料中含有呈色元素，以及烧窑后期人们能够控制火焰。陶器是一种质地较粗，具有吸水性、不透明的黏土制品。可视为人类创造的第一种新物质。

陶不仅是一种用于制造日常用品的物质，同时也适用于制造砖、瓦以及艺术品，典型的例子就是秦兵马俑。

China is one of the earliest countries in the world to have invented pottery; and its history of ceramic production can be traced back to some 7,000 to 8,000 years ago. During the Neolithic Age, there existed a great variety of pottery wares, such as grey pottery, red pottery, painted pottery, black pottery, white pottery and glazed pottery… to name just a few. They show different colors mainly because most of the raw materials used for their production contain color-presenting elements; and in the later stage of firing, people could control the flames in the kilns. As a clay product, pottery is of coarse texture, water absorbent and opaque; it can be regarded as the first manmade new material.

Pottery was not only a material used for making household utensils but also a substance equally suited to the manufacturing of bricks, tiles or works of art, an example in point is the life-sized Qin Dynasty terra-cotta warriors and horses.

1 灰陶 *Gray Pottery*

在陶器接近烧成时，由窑顶部灌水封顶，冷却后陶器呈灰色，称为灰陶。商至战国时期灰陶最为盛行。

During the kilning process, water was poured into the kiln from its top for cooling purpose. That will result what we called grey pottery. Grey pottery was extremely popular during the Shang Dynasty and the Warring States Period.

2 红陶
Red Pottery

　　红陶是在新石器时代早期，人们采用露天堆烧法或原始窑烧制成的陶器。因为在烧制过程中与大量空气接触，所以烧制成的陶器为红色，故称为红陶。

In the early part of the Neolithic Age, the firing of pottery took place in the open-air or inside primitive kilns. During the firing process, the baked pottery had contact with a huge amount of air; the end product became reddish brown, hence the term of "red pottery".

3 彩陶
Painted Pottery

　　彩陶是在陶坯表面用红、白、黑等矿物颜料绘有人物、动物、几何纹形的陶器。入窑烧制后花纹即附着于器物表面，不易脱落，称为彩陶。彩陶主要用料为三氧化二铁、锰铁矿、方解石、硬石膏等。

As its name suggests, painted pottery refers to pottery wares with painted designs of human figures, animals, geometric patterns and so on. Usually, the painting was done with red, white and black mineral pigments. After the kilning process, the painted designs would firmly adhere to the surface of the wares. In the ancient times, the major raw materials used in making painted pottery included: ferric oxide, ferromanganese, calcite, anhydrite etc.

4 黑陶
Black Pottery

用含铁量5%～9%的漂积土，经过精细淘洗、陈腐后制坯，装入匣钵，再经900～1050摄氏度高温烧制出的陶器，称为黑陶。

Black pottery was made from "rinsing clay" with iron content between 5% and 9%. After being carefully washed and staled, the clay was made into a piece of greenware and put into a sagger for firing at a high temperature of 900°C to 1050°C.

5 白陶
White Pottery

用含铁量2%以下的陶土经1000摄氏度左右的高温烧制。在还原氧的氛围中烧制出的陶器为白色，称为白陶。

If the greenware of a pottery piece was made from clay whose iron content was lower than 2%, and then fired at a high temperature of about 1,000°C in the oxygen reduction atmosphere, white color would appear on the pottery. Potter produced in this manner is called "white pottery".

瓷器 Porcelain

　　瓷器是古人在长期的制陶生产实践中，随着技术的不断提升和原材料选择的不断扩展而获得的一种发明。

　　从陶器到瓷器的跨越经历了非常漫长的历史过程。原始青瓷最早出现于商代初期，春秋时期进一步发展和完善。东汉中后期，浙江出产的青瓷已具备了真正的瓷器所应有的全部特性，这也标志着古人创造瓷器的过程的完结。这一成就应该归功于三个重大突破：采用瓷土（高岭土）作原料，釉的发明，为获取高温不断改进窑炉。

　　与陶器相比，瓷器质地更加致密，呈半透明状，不透水，硬度大，光滑，壁薄，颜色和形状也更加丰富，叩之则声音如磬。

Porcelain was not actually an invention. It was more a discovery by the ancient Chinese as a result of their millenniums of experimentation with more and more sophisticated ceramic-making techniques and raw materials.

The leap from pottery to porcelain was an extremely long historical process. Proto-celadon first appeared in the early part of the Shang Dynasty; it witnessed further improvements during the Spring and Autumn Period. By the mid to late period of the Eastern Han Dynasty, celadon produced in the southern province of Zhejiang possessed almost all the properties of real porcelain, marking the completion of the process of porcelain creation, which was only made possible thanks to three breakthroughs: the use of porcelain clay (kaolin clay); the invention of glaze; and the improvements of kiln furnaces to achieve high temperatures.

Compared with pottery wares, porcelains have close texture; therefore they are not only translucent and waterproofing but also are harder, smoother and thinner-walled, with a broader spectrum of colors and shapes. When knocked, they can produce the resonance of a chime-stone.

1 原始青瓷 *Proto-Celadon*

商代是从陶器过渡到瓷器的渐进阶段，也是原始青瓷发展的阶段。原始青瓷使用的釉料是用石灰石加黏土配制而成的，含铁元素，在氧化过程中烧成，烧成温度在 1000 摄氏度以上，呈青绿色。但它在原料加工、器物成型以及烧窑技术上还比较原始，故称"原始青瓷"。

The transition of pottery to porcelain started to take place step by step during the Shang and Zhou Dynasties. This duration is also the development period of proto-celadon. Using the kind of iron-containing glaze that was made from limestone and clay, proto-celadon, sometimes referred to as "glazed pottery", was fired in the oxidation process at temperatures of over 1,000°C, resulting in a bluish green color. The term "proto-celadon" indicates the relatively primitive working procedure involving raw materials processing, clay molding and kilning.

到 10 世纪的唐五代时期，烧窑技术的改进使细白瓷的生产成为可能。白瓷被誉为"人造玉石"。是以含铁量低的瓷坯，施以纯净的透明釉烧制而成的瓷器。白瓷创烧于东汉，到隋代才普及。白瓷也是画作及烧制五彩瓷、斗彩瓷、青花瓷最好的底背基础瓷。

By the 10th century during the Tang and Five Dynasties periods, improved kilning techniques had made possible the production of fine white porcelain. Known as "artificial jade", white porcelain is a ceramic made of porcelain clay with low iron content and coated with pure transparent glaze. It first appeared in the Eastern Han Dynasty and became prevalent during the Sui Dynasty. The many varieties of porcelain, such as "Five Colors", "Contesting Colors" and blue-and-white porcelains, were all made based on white porcelain.

元代将一定量的含铜物质作为着色剂，掺入釉料中，经烧制后呈现红色，故称"红釉瓷器"。

Originating from the Yuan Dynasty, red glazed porcelains were made using a special kind of glaze mixed with a substance containing copper coloring pigment. The glaze applied on the body of the porcelain piece would turn red after firing, hence the term "red-glazed porcelain".

青花瓷始于唐代。用天然钴料作颜料，在瓷坯上绘画或印图，后涂透明釉（这种工艺称作"釉下彩"），经高温一次烧成。

The origins of blue-and-white porcelain can be traced back to the Tang Dynasty. With natural cobalt as pigment, patterns were first painted or imprinted on the greenware before it was coated with a kind of transparent glaze; and this process was called underglaze decoration. After a one-time firing at high temperatures, a blue-and-white porcelain piece was thus produced.

5 斗彩
Porcelain of "Contesting Colors"

斗彩创烧于明代。斗彩的烧制过程如下：在瓷坯上用青花料勾绘出轮廓线，施釉后经高温烧制，冷却后再于轮廓线内填以彩绘颜料，经低温再次烘烧而成。由于釉下青花与釉上彩绘争奇斗艳，故名为"斗彩"。

Initiated in the Ming Dynasty, the process of making "Contesting Colors" porcelain involved several steps: first, to outline the patterns on the greenware with natural cobalt; second, to coat the patterned greenware with glaze; and third, to put it into the kiln for firing at high temperatures. After cooling, the outlined patterns on the porcelain piece were painted with pigments of different colors and then kilned again, this time at low temperatures. For the finished porcelain piece, the blue and white patterns under the glaze and the colorful paintings above the glaze contested with each other for beauty. Traditionally, porcelains produced in this manner were called "Contesting Colors".

　　唐三彩是一种低温釉陶器，因多以黄、褐、绿三色为主而得名。在色釉中加入不同的金属氧化物，经过焙烧，形成浅黄、赭黄、浅绿、深绿、天蓝、褐红、茄紫等多种色彩。

　　除了具有丰富的色彩外，由于唐代时与国外交往频繁，故唐三彩在自身相对简单、传统的形制基础上，又融入了异国元素。

A kind of glazed pottery produced at low temperatures, the tri-colored glazed Tang wares gained its name from its predominant colors of yellow, brown and green. In its production process, different metal oxides were added into the glaze, which, after being fired, would result in the greater richness of colors such as light yellow, brownish yellow, light green, dark green, blue, brownish red and purple etc.

Besides richer colors, exotic elements were also added to the originally simple, classic Chinese shapes of the tri-colored glazed Tang wares as a result of China's extensive contacts with foreign countries during the Tang Dynasty.

五彩瓷器简称"五彩"，始于明代，以红、黄、蓝、绿、紫彩料颜色为基础，在烧好的白瓷胎上绘画，再经低温烧制而成。五彩瓷有时又被称作釉上五彩瓷。

Originating from the Ming Dynasty, the so called "Five-Colored Porcelains", or simply "Five Colors", were made based on the pigments of five colors, namely, red, yellow, blue, green and purple. In the process of production, painting was first done on the fired white porcelain base. Afterward, the painted porcelain base was fired again at low temperatures. Five-colored porcelains are sometimes referred to as "overglazed five-colored porcelains".

珐琅彩瓷器由景泰蓝演变而来，在瓷胎上使用珐琅彩装饰手法制造。它起始于清康熙后期，专供宫廷皇室玩赏之用。

Evolving from cloisonné, "Cloisonné Colors" were made using the enameling technique, which allowed patterns on the finished porcelain pieces to be refined still further. They originated from the late period of Emperor Kangxi's reign of the Qing Dynasty and were used as curios exclusively for royal appreciation.

　　粉彩瓷器是釉上彩品种之一，始于清康熙年间，盛行于清雍正年间。制作过程如下：在烧好的白瓷胎上用"玻璃白"打底，在上面染彩作画，经低温烧制而成。其主要特点是改变了"五彩"单线平涂的色调，色彩层次更加丰富。

A kind of overglazed porcelain originating from Emperor Kangxi's reign and becoming fashionable during Emperor Yongzheng's reign of the Qing Dynasty, "famille rose" was produced in this manner: firstly, the kilned white porcelain base was coated with a kind of transparent glaze called "glass white"; secondly, coloring and painting was done upon the glaze; thirdly, the piece was fired again at low temperatures. In contrast to the single line flat coating technique applied in the production of "Five Colors", "famille rose" featured more gradations and greater richness of colors.

宋代五大名窑及八大窑系 *The Five Most Famous Kilns & the Eight Best Known Kiln Series of the Song Dynasty*

宋代是我国瓷业发展史上的一个繁荣时期，瓷窑遍布全国各地，名窑蜂起、名瓷迭出。宋代瓷窑各具特色，其中以五大名窑、八大窑系为代表，最能突显沉静素雅的美学风格。

五大名窑指官窑、哥窑、汝窑、定窑和钧窑。八大窑系指定窑系、磁州窑系、耀州窑系、钧窑系、龙泉窑系、景德镇窑系、建窑系和越窑系。

The Song Dynasty witnessed a flourishing historical period in terms of the development of the Chinese ceramic industry. During this period, kilns were built all over the country and famous porcelain pieces kept emerging. While Song porcelains were of diverse characteristics, the pieces produced by "the five famous kilns" and "the eight best-known kiln series" were the most representative of the Song people's aesthetic style of being serene, simple and elegant.

"The five most famous kilns" refer to the Official Kilns, the Ge Kilns, the Ru Kilns, the Ding Kilns and the Jun Kilns; "the eight best-known kiln series" refer to the Ding Kiln Series, the Cizhou Kiln Series, the Yaozhou Kiln Series, the Jun Kiln Series, the Longquan Kiln Series, the Jingdezhen Kiln Series, the Jian Kiln Series and the Yue Kiln Series.

1 定窑系 *The Ding Kiln Series*

定窑系以河北省曲阳涧瓷村及东西燕山村瓷窑为代表。曲阳县旧时称定州，故得此名。烧瓷始于唐代，盛于宋代，元代以后逐渐衰落，以烧民间用白瓷为主，兼烧黑瓷、酱釉和绿釉瓷。碗、盘类多用覆烧法，所以口沿部位多无釉，称为"芒口"。瓷器外部有薄层垂釉实为定窑之特征。定窑在北宋时期一度为宫廷烧制御用瓷。

The Ding Kiln Series was represented by the porcelain kilns built in Jianci Village and the western and eastern Yanshan villages of Quyang County in Hebei Province. It got its name as Quyang County had traditionally been called "Dingzhou". The ceramic production at Quyang County began in the Tang Dynasty, flourished during the Song Dynasty and gradually declined after the Yuan Dynasty. Thereafter, the production was mainly focused on white porcelains for civilian use, while black porcelains and porcelains in reddish brown and green color glaze were also made. Since the utilitarian pieces produced by Ding Kilns, such as bowls and plates were made with the so-called covered kilning method, the rim of the porcelain made in this manner was in most cases without glaze. Therefore, it was described as "awn mouth". One unique feature of the porcelain products of the Ding Kilns Series was a thin layer of drooping glaze on their outer surfaces. During the Northern Song Dynasty, porcelains commissioned by the imperial court were once made for royal use.

2 磁州窑系
The Cizhou Kiln Series

磁州窑系以河北省磁县观台镇为中心的瓷窑为代表，是中国北方最大的民窑体系。磁州窑系主产民间用瓷，品种丰富，既有白、黑单色釉瓷，又有白色绘花、刻花、剔花等瓷。图饰刻绘生动活泼，大都为民众所喜爱的花、鸟、鱼、虫、婴戏等。其中白色釉黑花最有特色，也最受广大民众喜爱。

Represented by the porcelain kilns built in Guantai Town of Cixian County in Hebei Province, the Cizhou Kiln Series was the biggest civil kiln system in northern China. It mainly produced porcelains for civilian use, which included both porcelains in monochromatic glaze like white and black, as well as porcelains vividly decorated with painted, engraved or carved designs in white color, mostly in the popular forms of flowers, birds, fish, insects and playing kids. Of the many varieties of porcelains produced by the Cizhou potteries, the pieces with black patterns on the white glazed background were considered the most characteristic and hence most favored by the people.

3 耀州窑系
The Yaozhou Kiln Series

　　耀州窑系以今陕西省铜川市黄堡镇的瓷窑为代表。宋时属耀州，故得此名。耀州窑系兴盛时期为宋代，主要烧制青瓷及彩釉，兼烧黑釉及白瓷。耀州窑青瓷多以刻花、印花装饰。在宋代诸多同类产品中，以耀州窑刻花瓷器为上乘。

Represented by the porcelain kilns built in Huangbao Town of Tongchuan City (called Yaozhou in the Song Dynasty, hence the name of the series) in Shaanxi Province, the Yaozhou Kiln Series was at its peak during the Song Dynasty. The kilns there mainly produced celadon and colored glazed porcelains; they also produced black glazed porcelains and white porcelains. The celadon pieces produced by the Yaozhou potteries generally featured engraved or imprinted decorations; and of the many varieties of similar products in the Song Dynasty, the ones with engraved designs produced by the Yaozhou kilns were considered the most distinguished.

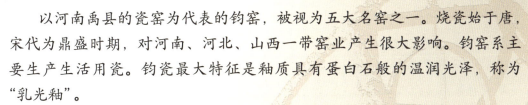

以河南禹县的瓷窑为代表的钧窑，被视为五大名窑之一。烧瓷始于唐，宋代为鼎盛时期，对河南、河北、山西一带窑业产生很大影响。钧窑系主要生产生活用瓷。钧瓷最大特征是釉质具有蛋白石般的温润光泽，称为"乳光釉"。

宋代钧窑首创用铜的氧化物为着色剂，在还原气氛中烧成铜红釉，为我国陶瓷工艺开创了一个新境界。

Represented by the porcelain kilns built in Yuxian County of Henan Province, the Jun Kiln Series is considered one of the five most famous kilns. Ceramic production there began from the Tang Dynasty and reached its heyday during the Song Dynasty, having a great impact on the ceramic production in areas in today's Henan, Hebei and Shanxi provinces. The major products of the Jun Kiln Series were utilitarian porcelains, characterized by their gentle opal luster resulting from the quality of the "shinning milky glaze".

During the Song Dynasty, the potteries of the Jun Kiln Series were the first to use oxides of copper as coloring materials for their products, which were fired in reducing atmosphere to become copper red glazed porcelains. This technique has helped open up a new boundary for China's ceramic technology.

5 龙泉窑系
The Longquan Kiln Series

龙泉窑系以浙江省龙泉县瓷窑为代表，以烧造青瓷闻名于世，兴盛于南宋。龙泉窑不同于其他窑所使用的石灰质釉，工匠们自己发明了石灰碱釉，使其釉面薄且晶莹剔透，胎质灰白。龙泉窑系主要生产盆、钵、瓶等物，其最具特色的是粉青釉及梅子青釉。

Represented by the so-called "dragon kilns" built on shallow slopes in Longquan County of Zhejiang Province, the Longquan Kiln Series has gained its international fame for its celadon pieces. Its flourishing time was in the Southern Song Dynasty. Unlike other kilns which used glaze made from limestone, the Longquan Kiln Series invented its unique lime alkali glaze. As a result, its products were characterized by their thin and crystal-clear glaze layer over the grayish white roughcast. Household utensils such as basins, bowls and vases constituted the major products of the Longquan Kiln Series, and they were best known for their pink blue glaze and plum blue glaze.

景德镇窑系源于6世纪南北朝时期江西景德镇瓷窑，既生产生活用瓷，也生产艺术瓷。除了为数众多的民用瓷窑生产民用瓷器外，还有御窑用来生产皇家贵族用瓷。在景德镇出产的种类繁多的瓷器中，最为出名的是宋代的"影青"、元代的"釉下红"和"青花"，以及明清的"斗彩""粉彩"和"五彩"。时至今日，景德镇仍然是中国制瓷业的代名词。

Originating from the porcelain kilns built in Jingdezhen of Jiangxi Province in the 6th century during the Northern and Southern Dynasties, the Jingdezhen Kiln Series produced both utilitarian and artistic porcelains. In addition to the numerous civil kilns that turned out products for civilian use, imperial kilns were also set up there to produce porcelains for use in the royal household and in those of the nobility. Most famous among the great variety of the Jingdezhen porcelain products were the "Shadowy Blue" of the Song Dynasty, the "Underglazed Red" and the "Blue-and-White" of the Yuan Dynasty, as well as the "Contesting Colors", "Famille Rose" and "Five Colors" of the Ming and Qing Dynasties. Even today, Jingdezhen is still synonymous with Chinese porcelain manufacture.

建窑系以福建省建阳县瓷窑为代表，又称"建阳窑"，主烧黑釉茶盏。其胎色深黑坚硬，俗称"铁胎"。盏内外均施黑釉。器物外釉不到底，露胎。以铁结晶形成的斑纹为饰。

Represented by the porcelain kilns built in Jianyang County of Fujian Province, the Jian Kiln Series is also referred to as the Jianyang Kiln Series. Its main products were black glazed teacups. As the color of the roughcast of the pieces was solid and in dark black, the products of the Jian Kiln Series were vividly called "iron roughcast". Black glaze was applied to both the inside and the outside of the pieces except their lower surface where the roughcast was exposed. The products often featured streaks resulting from iron crystallization.

越窑系以浙江省余姚上林湖滨地区的瓷窑为代表，因地处越国故地而得名。宋代瓷器主供宫廷。其工艺精湛，造型工整，釉色青绿。该窑后来生产了大量民间用瓷。

Originating from the Yue kilns built in areas around the Shanglin Lake of Yuyao in Zhejiang Province, the Yue Kiln Series is named after the ancient Kingdom of Yue. In the Song Dynasty, it mainly produced porcelains for the exclusive use of the royal household; and its products were noted for their perfect craftsmanship, neat shapes and dark green glaze. From the Song Dynasty onwards, it began to mass produce porcelain wares for civilian use.

中国历史年表 Chinese Chronology		
朝代 **Dynasty**		**年代** **Age**
夏 Xia Dynasty		前 2070—前 1600 2070 B.C.—1600 B.C.
商 Shang Dynasty		前 1600—前 1046 1600 B.C.—1046 B.C.
周 Zhou Dynasty	西周 Western Zhou Dynasty	前 1046—前 771 1046 B.C.—771 B.C.
	东周 Eastern Zhou Dynasty*	前 770—前 256 770 B.C.—256 B.C.
秦 Qin Dynasty		前 221—前 206 221 B.C.—206 B.C.
汉 Han Dynasty	西汉 Western Han Dynasty	前 206—公元 25 206 B.C.—A.D. 25
	东汉 Eastern Han Dynasty	25—220 A.D. 25—A.D. 220
三国 Three Kingdoms Period		220—280 A.D. 220—A.D. 280
西晋 东晋 Jin Dynasty		265—420 A.D. 265—A.D. 439
南北朝 Northern and Southern Dynasties	南朝 Southern Dynasty	420—589 A.D. 420—A.D. 589
	北朝 Northern Dynasty	386—581 A.D. 386—A.D. 581
隋 Sui Dynasty		581—618 A.D. 581—A.D. 618
唐 Tang Dynasty		618—907 A.D. 618—A.D. 907
五代 Five Dynasties		907—960 A.D. 907—A.D. 960
宋 Song Dynasty	北宋 Northern Song Dynasty	960—1127 A.D. 960—A.D. 1127
	南宋 Southern Song Dynasty	1127—1279 A.D. 1127—A.D. 1279
元 Yuan Dynasty		1206—1368 A.D. 1206—A.D. 1368
明 Ming Dynasty		1368—1644 A.D. 1368—A.D. 1644
清 Qing Dynasty		1616—1911 A.D. 1616—A.D. 1911

* 春秋时代：前 770—前 476（The Sping And Autumn Period: 770B.C.—476B.C.）

战国时代：前 475—前 221（The Warring State Period: 475B.C.—221B.C.）